Manures, Fertilizers, and Soil Fertility Management

NIPA® GENX ELECTRONIC RESOURCES & SOLUTIONS P. LTD.
New Delhi-110 034

About the Authors

Dr. R.K. Gupta graduated in 1983, specializing in Soil Science for his Master degree in 1985 and earning his Ph. D as an in-service candidate in 1998 from PAU, Ludhiana. Dr Gupta started his professional career as an Assistant Soil Scientist in 1986 and was elevated to the position of Soil Chemist and then to Principal Soil Chemist in 2006 and retired as Head of the Department of Soil Science, Punjab Agricultural University, Ludhiana. Dr. Gupta has made outstanding contributions to Integrated Nutrient Management by substantiating fertilizer requirements by working out N and P contributions from biogas-slurry, poultry manure, press mud, crop residues, and biochar in rice-wheat cropping systems. Much of his research was devoted to increasing fertilizer N use efficiency in rice, wheat, maize and cotton crops through neem-coated urea, optimization, and synchronization of N application with crop N demand using a leaf colour chart, SPAD meter and Green Seeker.

Dr Gupta developed indigenous technology for biochar manufacturing from rice residues as an alternative to farmers' burning rice residue. This technology has been adopted by 5 KVKs in Punjab and 11 KVKs in Bihar and is now commercialized by the Punjab Agricultural University. His work on crop residue management in rice-wheat cropping systems, either by incorporating it into the soil or retaining it on the soil surface, helped in yield advantage and saved 20 kg N ha^{-1} in wheat after three years over residue removal or burning. He developed more than 45 practical recommendations for improving soil health, saving fertilizer use, and increasing farmers' profits. Besides, he has 100 original research articles, one review paper, four books, seven book chapters, four research bulletins, and 155 extension articles. He has handled 14 national/international projects as PI/Co-PI. He has mentored 15 PG students as a Major advisor. Under his guidance, All India coordinated research project on STCR at PAU Ludhiana and awarded Ch. Devi Lal ICAR award in 2010. He also received the FAI Golden Jubilee Award for Excellence for the best work done in Plant Nutrition in 2020 and the Prof. Manjeet S. Chinnan Distinguished Professor Chair Award in 2022 by PAU, Ludhiana. He was also awarded an appreciation letter from the Department of Agriculture, Govt. of Bihar, for providing guidance and training in producing biochar from rice straw in 2022. He is currently a Professor at Lovely Professional University, Jalandhar.

Dr. Sohan Singh Walia, Director-cum-Principal Agronomist, School of Organic Farming, Punjab Agricultural University, Ludhiana has joined PAU as Assistant Professor (Agronomy) in 1995, elevated to the position of Principal Agronomist in 2010. He started organic farming research as a pioneer work at PAU, Ludhiana as Ph.D. student (2001-04) and then worked on organic farming and integrated farming systems in All India Co-ordinated Research Project on Integrated Farming Systems and All India Network Project on Organic Farming. Dr Walia has to his credit more than 450 research and extension publications, nine books, ten teaching manuals, seven extension folders and 22 book chapters. He has handled 24 research projects and is currently handling five research projects. The chapter on organic farming and integrated farming system in the package of practices was included during 2004-05 and 2017-18, respectively. Ninety-seven recommendations have been included in the package of practices for mass adoption by the Punjab Farmers especially, resource conservative cropping systems, nine organic farming-based cropping systems and production technology of cultivation of direct seeded rice. In addition, he was involved in technologies regarding the application of consortiums in sugarcane, turmeric, potato, onion, maize, and wheat crops; integrated nutrient management in maize/soybean, rice residue management for the mass scale adoption under Punjab conditions. Developed Integrated Farming System Research Model comprising dairy, fishery, horticulture, vegetables, agro-forestry, vermicomposting. He has been involved in the teaching of 93 courses. Dr Walia has guided four Ph.D. and eleven M.Sc. students as major advisor. He has organized 20 training programmes, delivered 62 invited lectures, 350 training lectures, and 52 TV and radio talks. He has attended ten international and 92 national conferences/seminars/workshops.

Dr Walia was appreciated for outstanding work on integrated (2007-2017) and organic farming during QRT review (2012-2017). Recipient of Best organic farming centre award (2019) by ICAR; Dr M S Randhawa best book award (2017); Fellow, Indian Ecological Society (2016); Gold medal from Society of Recent Development in Agriculture at International Conference (2013); ISA Best Paper Award along with cash prize of Rs 5000/- from Indian Society of Agronomy (2011) for paper entitled, Alternate cropping systems to rice-wheat for Punjab"; ISA P. S. Deshmukh Young Agronomist Award (2005) by Indian Society of Agronomy; Fellow, Society of Environmental Sciences (2004). The AICRP on Integrated Farming System, Ludhiana centre received the Best Centre Award (2014 and 2022) from ICAR- Indian Institute of Farming Systems Research, Modipuram, Meerut. Received Dr. Gurbaksh Singh Gill Gold Medal, Merit Certificate for M.Sc. (Agronomy); S.S. Labh Singh Gold Medal and Merit Certificate for B.Sc. Agri. (Hons.). Dr Walia has acquired advance training in rice production systems from School of Agriculture, Food and Wine, University of Adelaide, South Australia.

Miss. Tamanpreet Kaur has completed her Masters from Punjab Agricultural University, Ludhiana. She is working as Senior Research Fellow (2018–till date) under project entitled "All India Co-ordinated Research Project on Integrated Farming Systems" at School of Organic Farming, Punjab Agricultural University, Ludhiana. He holds a M.Sc. in Agriculture from Punjab Agricultural University, Ludhiana. Her area of interest is on integrated farming systems. She has eight research papers, 13 extension articles, three booklet on integrated farming systems, and three book chapters.

Manures, Fertilizers, and Soil Fertility Management

R.K. Gupta
Principal Soil Chemist
School of Agriculture, Lovely Professional University
Phagwara, Punjab
Former Head, Department of Soil Science
Punjab Agricultural University
Ludhiana, Punjab

S.S. Walia
Director-cum-Principal Agronomist
School of Organic Farming,
Punjab Agricultural University
Ludhiana, Punjab

Tamanpreet Kaur
Senior Research Fellow
School of Organic Farming
Punjab Agricultural University
Ludhiana, Punjab

NIPA® GENX ELECTRONIC RESOURCES & SOLUTIONS P. LTD.
New Delhi-110 034

NIPA® GENX ELECTRONIC RESOURCES & SOLUTIONS P. LTD.

101,103, Vikas Surya Plaza, CU Block
L.S.C. Market, Pitam Pura, New Delhi-110 034
Ph : +91-11-43860225, Mob.: +91 9717133558, 9540816132
E-mail: newindiapublishingagency@gmail.com
Website: www.nipaersources.com

Print ISBN: 978-93-5887-181-4

ebook ISBN: 978-93-5887-485-3

Composed and Designed by NIPA®.

Preface

The sustainable management of soil fertility is central to achieving food security and ecological stability in modern agriculture. With the increasing challenges posed by population growth, diminishing natural resources, and climate change, a scientific understanding of soil nutrient management has become more critical than ever. This book, "Manures, Fertilizers, and Soil Fertility Management", is a comprehensive guide designed to provide readers with theoretical insights and practical tools for managing soil fertility effectively and sustainably.

The book begins with an exploration of the history of soil fertility and plant nutrition, presenting the evolution of scientific understanding in this field. Fundamental concepts such as essential and beneficial elements, criteria of essentiality, and the roles of essential plant nutrients are discussed, along with methods for identifying nutrient deficiencies through plant symptoms. A detailed examination of macro and micronutrients—including nitrogen, phosphorus, potassium, secondary nutrients, and micronutrients—provides a robust foundation for understanding their functions, dynamics, and management. A significant focus is placed on soil fertility evaluation, highlighting methods for assessing soil nutrient status through laboratory analysis and field diagnostics. The book also delves into fertilizer use efficiency and the methods and timing of fertilizer application, ensuring readers gain practical knowledge for optimizing nutrient uptake by crops.

To address the growing interest in sustainable agriculture, this book explores organic manures, chemical fertilizers and their classification, and nanofertilizers, with an emphasis on their properties, applications, and integration within modern nutrient management systems. The Fertilizer Control Order and the concept of Integrated Nutrient Management (INM) are also covered to ensure readers are acquainted with regulatory and sustainable practices. Additionally, the book discusses soil amendments, providing insights into improving soil structure and correcting soil pH for better crop performance.

A unique feature of this book is its dedicated practical section, offering step-by-step procedures for the analysis of soils and plant samples. Techniques include the determination of available nutrients in soils (e.g., phosphorus, potassium, sulphur, and micronutrients), exchangeable cations, and organic carbon. Readers will also find practical methods for analyzing total nutrients in plant samples, along with an introduction to the principles and applications of analytical instruments used in soil and plant testing laboratories.

This book is intended for students, researchers, agricultural professionals, and policymakers seeking an in-depth understanding of soil fertility management. Its integration of theoretical concepts with practical methodologies makes it a valuable resource for both academic and applied purposes.

We are confident that this book will serve as a useful reference for anyone committed to improving agricultural productivity while safeguarding soil health. Feedback and suggestions from readers are welcome to refine and enhance the content for future editions.

Authors

Contents

Practicals

Theoretical

1

Soil Fertility History and Plant Nutrition

The active management of soil nutrients to ensure optimal plant nutrition has been a fundamental practice in agriculture since its inception. For centuries, production of agriculture depended on the organic materials recycling like manure and crop residues. Interest in plant nutrition dates back to ancient times, first captivating philosophers and later becoming a subject of rigorous scientific inquiry. The history of how plants obtain nutrients is marked by numerous misconceptions and incorrect theories. Exploring the long-standing efforts to uncover this essential agricultural process highlights the perseverance and brilliance of early thinkers and scientists. It also underscores that modern understanding, which may now seem straightforward, is the culmination of significant scientific endeavour. The concept of managing fertility of soil to boost yield of crops is far from new—its origins are intertwined with the earliest practices of agriculture. Ancient civilizations recognized the value of applying organic matter such as green manure, farmyard manure, night soil, bone, and wood ashes to enhance soil productivity.

Ancient History

- Ancient Egyptian murals depict various agricultural activities such as tillage, planting, irrigation, and harvesting, yet they do not illustrate any fertilization practices.
- Roman art provides significant insights into their agricultural practices. Alongside Saturn and Ceres, the Roman Gods of agriculture, the Romans also worshiped Stercutius, the god of manure, who women and children particularly venerated.
- Several Roman writers, such as Cato, Columella, and Pliny the Elder, authored comprehensive texts on agriculture. These works discussed the application of animal manures, green manures, and strategies to enhance soil productivity.
- Pliny the Elder (AD 23-79) categorized various animal manures, noting their respective advantages and disadvantages. He also advocated for the use of green manures, stating, "All authors concur that incorporating a crop of lupines into the soil before pod formation—whether through ploughing or forking or cutting and burying them in mounds around the bases of trees and vines is highly advantageous."

- The earliest documented use of green manures for soil improvement dates back to the Chou dynasty (~1000 BC) in China. By 500 BC, Tsi stated that "Green manures are sown in the fifth or sixth month and incorporated into the soil in the seventh or eighth month and its fertilizing value is comparable to silkworm excrement and well-decomposed farm manure."
- FH King (1911) in his book "Farmers of Forty Centuries" stated that "We sought to understand how, after twenty, thirty, or even forty centuries, their soils have been able to sustain such large populations."
- Chinampas agriculture in Central America involved the creation of actual floating gardens, allowing for the development of cultivation areas in the shallow regions of lakes.

Middle Age

In 1240, Crescentius compiled and summarized Roman agricultural literature. Bernard Palissy (1510-1589), a ceramicist and hydraulic engineer, wrote: "Manure is applied to the fields to return a portion of what was taken away. By doing so, the nutrients which were removed by previous crops are restored to the soil, which future crops will then utilize to their benefit."

Early Modern Period (16th–19th Century)

- Van Helmont's (1577-1644) willow tree experiment marked a significant milestone in the understanding of plant nutrition. He explored the concept of vegetation and investigated the question of what plants consume for their growth.
- Van Helmont conducted experiments where he burned charcoal and observed that 62 pounds of charcoal produced only 1 pound of ash. He concluded that the remaining 61 pounds were made up of the "spirit of the wood," which he referred to as "gas."
- Jethro Tull (1731) invented the grain drill, along with several other technologies that contributed to significant increases in grain yields during the 18th century.
- Tull's theory of plant nutrition suggested that all types of manure and compost contain substances that, when combined with soil, ferment and cause the soil to break down, crumble, and divide. According to Tull, the primary purpose of manure is not to directly nourish plants but to aid in breaking up the soil, making it more accessible for plant roots to absorb nutrients.
- *The Humus Theory*
  a. Plants, like animals, must consume organic substances similar in composition to themselves.

b. Soil organic matter, known as humus, is the primary nutrient source for plants and the key to soil fertility.

c. Plant roots absorb humus and, combined with water, transform it into plant tissue.

- Nicolas-Théodore De Saussure (1804), a Swiss chemist and plant physiologist challenged many earlier concepts of plant nutrition and made several groundbreaking discoveries:
  a. Soil, rather than air, provides plants with nitrogen.
  b. Roots actively and selectively absorb water and salts.
  c. Plants only take in CO_2 and release O_2 when exposed to sunlight.
  d. All plant tissues, including roots and seeds, consume oxygen and emit CO_2.
  e. He did not entirely reject the Humus Theory.
- Carl Sprengel (1838) rejected the humus theory, concluding that the importance of humus has been greatly overstated, and the prevailing doctrine of humus is rife with contradictions.
- Justus von Liebig (1840), a renowned German chemist, was a pioneer in the study of plant mineral nutrition. As the first professor to incorporate laboratory methods in chemistry teaching, he is considered one of the most influential chemistry educators in history. His authority in the field was so respected that few questioned his views on mineral nutrition. Although Liebig did not originally propose the Law of the Minimum, he is credited with popularizing the concept.
- Liebig believed that the ash produced by burning a crop contained the ideal balance of nutrients needed to fertilize that same crop and advocated for a "check book" method of fertilization.
- Although some theories of Liebig lacked experimental proof, few were willing to publicly challenge them. For instance, he claimed, "Agricultural crops receive sufficient nitrogen from the atmosphere."
- Lawes and Gilbert (1843) established the first agricultural experiment station at Rothamsted and initiated the Broadbalk experiment, which has involved continuous wheat cultivation for over 160 years. This remains the longest-running agricultural field experiment in the world.
- JB Lawes established one of Europe's first fertilizer factories and faced the challenge of persuading farmers to adopt his fertilizers.
- The Broadbalk experiment demonstrated the benefits of nitrogen and phosphorus fertilizers, concluding that manure was not essential for achieving high crop yields. However, plots treated with both manure and fertilizer frequently yielded the highest results. The Broadbalk archive now holds more than 200,000 samples of hay, grain, and soil.

- Hopkins (1913) initiated the Morrow Plots at the University of Illinois, making it the oldest experiment in the United States.
- In his book, Hopkins asserted that a farmer should know the names of 10 essential elements of plant nutrition just as well as he knows the names of his 10 closest neighbours.
- He also emphasized that the farmer's true wealth does not lie in the land itself, but in the soil components that provide nutrition for plants.
- Hopkins et al., (1908) stated in his book "Soil Fertility and Permanent Agriculture" that legumes are capable of fixing their own nitrogen if they are associated with the proper bacteria.
- In the 19th century, farmers had few options for purchasing nutrients. Some turned to guano imported from South America, which produced significant yield increases across various crops. Guano is a highly concentrated source of nitrogen (~12%) and phosphorus (P_2O_5, ~12%), offering much higher nutrient levels than most other organic fertilizers.

19th century

- Fritz Haber developed a laboratory apparatus for synthesizing ammonia from hydrogen and nitrogen, with the catalytic reaction occurring within a large cylinder.
- The remarkable success of the "chemicalization" of agriculture, along with the adoption of Green Revolution technologies in the 1980s, provided the capacity to sustain the global population.
- Extensive research was initiated to investigate the impact of fertilization on the mineral content and key organic compounds (such as amino acids, proteins, vitamins, hormones, and enzymes). This research led to the discovery of two new fertilizer laws by French scientist André Voisin (1903-1964) in the 1960s: the Law of the Maximum and the Law of the Priority of Biological Quality.

Law of the Maximum: An excess of any available substance in the soil can reduce the effectiveness of other nutrients, ultimately resulting in lower yields.

Law of the Priority of Biological Quality: The primary objective of using fertilizers should be to enhance the biological quality of crops, which holds greater significance than merely increasing the yield.

Modern times, 20th century

- In modern agriculture, there is a remarkable collaboration and integration of scientific research from fields such as agricultural chemistry, chemical technology, biology, and plant mineral nutrition, among others, in practical applications.

- Advances in plant nutrition have led to various chemical innovations, often employed individually or in combination. For instance, genetic engineering has been used to enhance the absorption of micro and macronutrients by plants. A significant achievement in this field is the expression of ferritin, a protein that stores iron in a bioavailable form. Recombinant soybean ferritin has been incorporated into several grains, and pea ferritin has been successfully expressed in rice.

The key milestones in the history of plant nutrition are as follows:

- Van Helmont (1600): What can be utilized as nutrients for plant growth?
- Priestly (1772): Oxygen is released by plants from leaves into the atmosphere
- De Saussure (1804): In addition to carbon dioxide (CO_2), water also plays a role in the process of photosynthesis
- Thaer (1806): The first academy of agriculture was established in Moglin, Germany
- Thaer (after 1800): The humus theory for plant nutrition
- Bousssingault (1834): The importance of nitrogen in plant nutrition
- Liebig (1841): Theory of mineral plant nutrition
- Sprengel and Liebig (1842): Theory of minimum in agricultural chemistry and Liebig's Law of the Minimum
- Lawes (1842): Patent issued for the production of single superphosphate
- Lawes (1848): First superphosphate factory was established in Liverpool
- Hellriegel and Wilfarth (1886): Discovery of nitrogen fixation by Rhizobia
- Haber and Bosch (1911): Synthesis of ammonia
- Borlaug (1970): Green Revolution: Technological advancements enabling global food production
- Robert (1994): Precision Agriculture
- Various Scientists (after 2000): Innovative fertilizers

Numerous overlooked instances of early perceptions predate today's widely accepted principles, such as the concept of sustainability, which is now seen as essential for managing natural resources. The remarkable advances in plant nutrition observed in recent years are largely attributed to developments in fields like chemistry, chemical technology, physics, biology, and the physiology of mineral nutrition, as well as the adoption of modern analytical methods and precision research tools. These innovative approaches have transformed our ability to measure plant nutrition processes, allowing for more accurate conclusions and minimizing errors. The progress in the study of mineral nutrition in plants is also driven by the need to boost food production, with the goal of feeding an estimated 9 billion people by the middle of this century. This responsibility lies with farmers, biologists, and chemists, among others. As Arnold Fink, a prominent German agricultural chemist, once stated, "Scientific theories and hypotheses are useful in agriculture only if they are validated when tested against nature."

References

Carl, Sprengel (1828): Von den Substanzen der Ackerkrume und des Untergrundes. Journal für Technische und Ökonomische Chemie, 2, 243-474, and 3, 42-99, 313-352, and 397-421.

Hopkins, C. G. (1913) The Illinois System of Permanent Fertility. Circular No. 167 https://hdl.handle.net/2142/33252 (University of Illinois Agricultural Experiment Station).

Hopkins, C. G., Eckhardt, W. G. & Readhimer, J. E. (1908). Thirty Years of Crop rotations on the Common Prairie Soil of Illinois. Bulletin No. 125 https://www.ideals.illinois.edu/handle/2142/3031 (University of Illinois Agricultural Experiment Station).

Jethro Tull, The New Horse-Houghing Husbandry (1731), p. xiv.

Justus von Liebig (1840): Die Chemie in ihrer Andwendung auf Agrikultur und Physiologie. Vieweg und Söhne, Braunschweig.

King, F. H. (1911). Farmers of Forty Centuries or Permanent Agriculture in China, Korea and Japan (First UK published edition, 1927). London: Jonathan Cape.

Lawes, J. B. & Gilbert, J. H. Report of experiments on the growth of wheat for twenty years in succession on the same land. *J. R. Agricult. Soc. England* **25** (Part I and II): 93–185, Part I 449-501 Part II (1843).

Nicolas-Théodore De Saussure (1804). Recherches Chimiques Sur la Végétation, Nyon, Paris, 1804; Ann. Chim. [2] 225-244.

2

Essential and Beneficial Elements, Criteria of Essentiality

Introduction

A total of 109 elements have been discovered and incorporated into the periodic table. Many of these elements are found in the earth's mantle, crust, and soil, though their abundance varies. The chemical makeup of plants is, to some degree, influenced by the composition of the soil. However, since plants are active organisms, they exhibit selectivity in absorbing elements from the soil. Plants accumulate elements at different rates, absorbing some in large quantities, others in smaller amounts, and excluding most altogether. Before proceeding, it is important to familiarise with several key terms frequently used in this chapter.

Nutrient: A nutrient element is essential for an organism to complete its life cycle, and a deficiency in this element leads to specific deficiency symptoms. The negative effects can only be corrected by providing that particular element. Nutrient elements essential for plants may not be necessary for other organisms, and vice versa. Depending on availability, the nutrient levels in a plant can be deficient, sufficient, or toxic. A nutrient is considered deficient when its concentration is so low that it significantly limits growth and causes recognisable deficiency symptoms. If nutrient levels cause reduced growth without visible deficiency symptoms, they are termed insufficient. The nutrient content range in plants that supports optimal yield of crops is considered sufficient. When the concentration of a nutrient becomes high enough to significantly reduce growth, it is considered toxic.

Available Nutrient: Soil nutrients exist in various chemical forms that are often in a state of dynamic equilibrium, forming a nutrient pool from which plants can absorb them. The soil's available form of a nutrient is the fraction that, when it fluctuates, significantly influences crop yield and plant response (Bray and Kurtz, 1945). The available forms of the soil nutrients depends not only on their physical and chemical characteristics but also on the plant's ability to access them through its root system. Nutrients that are accessible to biological organisms are referred to as bioavailable nutrients. In the context of plant nutrition, this refers to the nutrient portion of the soil that plants can readily absorb and assimilate. Only a small fraction of the total nutrients in the soil is available for plant uptake.

Beneficial Elements: Beneficial elements are minerals that enhance plant growth but are not strictly essential, or they may be essential only for particular species of plant or under specific environmental conditions. Examples of beneficial elements include sodium, silicon, aluminium, selenium, cobalt, and vanadium.

Functional Nutrient: Any element that contributes to metabolism of plant, regardless of whether its role is specific or indispensable (Nicholas, 1961).

Trace Element: A trace element occurs in very low concentrations, typically less than one part per million (ppm) or even lower, in soil, plants, water, and other environments.

Tracer Element: A tracer element is a radioisotope or stable isotope used to follow its pathway within a system, allowing researchers to study the mechanisms of its interactions within that system.

Heavy Metal: A metal with a specific gravity greater than 5.0 or an atomic number above 20 is classified as a heavy metal. Essentially, any metal heavier than calcium falls into this category.

Nutrient Content: This term refers to the concentration or amount of a nutrient per unit weight of plant tissue. Nutrient content is typically expressed as a percentage (kg/100 kg or g/100 g) or in parts per million (ppm), equivalent to milligrams per kilogram of dry matter (mg/kg) or micrograms per gram of dry matter (μg/g). To convert the percentage of a nutrient to parts per million, multiply by a factor of 10,000 or 10^4.

Nutrient Accumulation: Nutrient accumulation refers to the storage of nutrients in specific parts or sections of a plant. Various plant parts accumulate nutrients to differing extents.

Nutrient Uptake: Nutrient uptake is the total amount of nutrients absorbed by growing crops from soil or other sources. It is determined by the amount of plant biomass produced and the nutrient concentration within the plant. For example, if the roots, straw, and grain of a wheat crop, producing dry matter yields of 2.5, 6, and 4 t/ha, respectively, contain 0.2%, 0.5%, and 1.5% nitrogen (N), then the nitrogen uptake can be calculated as follows:

N uptake by roots = 2.5 (t/ha) × 0.2 (kg/100 kg) = 0.005 t/ha = 5.0 kg/ha

N uptake by straw = 6.0 (t/ha) × 0.5 (kg/100 kg) = 0.03 t/ha = 30 kg/ha

N uptake by roots = 4.0 (t/ha) × 1.5 (kg/100 kg) = 0.06 t/ha = 60 kg/ha

Total uptake by wheat = 5 + 30 + 60 = 95 kg/ha

Total N uptake by the above ground portion = 30 + 60 = 90 kg/ha

Nutrient removal: Nutrient removal refers to the extraction of nutrients found in the harvested part of a crop.

Essential Nutrients

These are elements required by plants to survive. For an element to be considered essential, it must meet the following criteria, as outlined by Arnon and Stout (1939):

1. The plant cannot complete its vegetative or reproductive cycle without the element.
2. The deficiency of the element is specific and can only be rectified by providing that particular nutrient.
3. The element plays a direct role in the plant's nutrition, independent of any indirect effects it may have on microbial or chemical conditions in the soil or growing medium.

The seventeen essential nutrients required for plant growth include carbon (C), hydrogen (H), oxygen (O), nitrogen (N), phosphorus (P), potassium (K), calcium (Ca), magnesium (Mg), sulphur (S), iron (Fe), manganese (Mn), zinc (Zn), copper (Cu), boron (B), molybdenum (Mo), chlorine (Cl), and nickel (Ni). Nickel was added to the list of essential nutrients in 1987. These nutrients are classified into macronutrients and micronutrients based on the amount needed by plants. Detailed information regarding their discovery, discoverers, and other relevant details is provided in Table 1.

Table 1: Key information on plant nutrients

Nutrient	Discovery year	Essentiality discovered authors (Discoverer)	Plant unstable form	Average concentration in plant tissue	Relative no. of atoms compared to Mo as one
H	-	Since time immemorial	H_2O	6.0%	60,000,000
O	-	Since time immemorial	H_2O, O_2	45.0%	30,000,000
C	1800	Priestley et al.	CO_2	45.0%	30,000,000
N	1804	Theodore de Saussure	NO_3^-, NH_4	1.5%	1,000,000
K	1839	C. Sprengel	K^+	1.0%	400,000
Ca	1839	C. Sprengel	Ca^{2+}	0.5%	200,000
Mg	1839	C. Sprengel	Mg^{2+}	0.2%	100,000
P	1839	C. Sprengel	$H_2PO_4^-$, HPO_4^{2-}	0.2%	30,000
S	1860	Sachs and Knop	SO_4^{2-}	0.1%	30,000
Cl	1954	T.C. Broyer, A.B. Carlton, C.M. Johnson & P.R. Stout	Cl^-	100 mg/kg	3,000
Fe	1843	E. Gris	Fe^{2+}	100 mg/kg	2,000

Nutrient	Discovery year	Essentiality discovered authors (Discoverer)	Plant unstable form	Average concentration in plant tissue	Relative no. of atoms compared to Mo as one
B	1923	K. Warington	H_3BO_3, $H_2BO_3^-$, HBO_3^{2-}, BO_3^{3-}	20 mg/kg	2,000
Mn	1922	J.S. McHargue	Mn^{2+}	20 mg/kg	2,000
Zn	1926	A.L. Sommer, C.P. Lipman	Zn^{2+}	20 mg/kg	300
Cu	1931	A.L. Sommer, C.P. Lipman & G. McKinney	Cu^{2+}	6 mg/kg	100
Mo	1939	D.I. Arnon, P.R. Stout	MoO_4^{2-}	0.1 mg/kg	1
Ni	1987	P.H. Brown, R.M. Welch & E.E. Cary	Ni^{2+}	0.1 mg/kg	1-2

Source: Tisdale et al. (1997)

Macronutrients

Macronutrients, also known as major nutrients, are required by plants in large amounts, exceeding the need for elements like iron. These include carbon (C), hydrogen (H), oxygen (O), nitrogen (N), phosphorus (P), potassium (K), calcium (Ca), magnesium (Mg), and sulphur (S). Carbon, hydrogen, and oxygen make up 90 to 95% of a plant's dry weight and are derived from carbon dioxide (CO_2) and water. The remaining six macronutrients—N, P, K, Ca, Mg, and S—are further classified into primary and secondary nutrients.

Primary nutrients, such as nitrogen, phosphorus, and potassium are needed by plants in the greatest amounts. Their deficiencies are typically addressed through the use of commercial fertilizers, which contain these elements as key ingredients.

Secondary nutrients, including calcium (Ca), magnesium (Mg), and sulphur (S), are required in moderate amounts by plants. Deficiencies of these nutrients are often localized and can be unintentionally corrected through the use of fertilizers that supply primary nutrients. For example, single superphosphate (SSP), a phosphatic fertilizer, also provides calcium and sulphur, while ammonium sulphate, a nitrogen-based fertilizer, adds sulphur as well.

Micronutrients

Micronutrients are essential nutrients that plants need in small quantities in comparison to macronutrients, but their role is just as crucial. These include elements such as iron, copper, manganese, zinc, boron, molybdenum, chlorine, and nickel. The significance of micronutrients is highlighted by the fact that the absence of a single atom of molybdenum (Mo) can make millions of atoms of oxygen, hydrogen, carbon, and nitrogen (N) ineffective in terms of plant nutrition. Micronutrients are categorized based on how plants absorb them: cationic micronutrients (Cu, Fe, Mn, Zn, Ni) and anionic micronutrients (B, Cl, Mo).

Functions and Deficiency Symptoms of Nutrients

Macronutrients

1. *Nitrogen*

Nitrogen is primarily taken up by roots of plant in the form of nitrate (NO_3^-), and in the case of rice, also as ammonium (NH_4^+). In nitrogen-sufficient plants, its concentration typically ranges from 1 to 5 percent (10,000 to 50,000 ppm or mg/kg). This nutrient is often limited in soils and is a common cause of significant nutritional challenges. Nitrogen plays a critical role in plant cells, forming a major component of both the cell structure and cell walls, where it can constitute up to 5 percent. Additionally, nitrogen is found in various amounts within the cytoplasm and organelles, where it combines with elements like carbon, hydrogen, oxygen, phosphorus, and sulphur.

Functions

1. Nitrogen is a vital component of proteins, amino acids, nucleic acids, porphyrins, flavins, purines, pyrimidine nucleotides, flavin nucleotides, enzymes, coenzymes, and alkaloids. In the presence of sunlight, nitrogen containing chlorophyll facilitates the fixation of atmospheric CO_2 into carbohydrates. As a key part of nucleic acids, including Ribonucleic Acid (RNA) and Deoxyribonucleic Acid (DNA), nitrogen plays a crucial role in passing genetic information to offspring. Proteins, which are made of nitrogen, form structural components and are involved in catalysing biochemical reactions.
2. Nitrogen improves the quality of fodder and leafy vegetables.
3. Nitrogen fertilization enhances the quality of protein of food grains by increasing the proportion of amino acids such as proline, glutamic acid, phenylalanine, cystine, methionine, and tyrosine, while reducing lysine, arginine, histidine, aspartic acid, threonine, valine, glycine, and leucine in the grain.

Deficiency Symptoms

Plants with less than one percent nitrogen content are typically considered nitrogen deficient. Stunted growth is a common sign of nitrogen deficiency. Since nitrogen is highly mobile within the plant, deficiency symptoms usually first appear on older leaves, which turn light green to pale yellow due to proteolysis. The intensity of this discoloration depends on the severity of the deficiency. Nitrogen deficiency is also linked to reduced flowering, lower crop yields, and decreased protein content.

Toxicity symptoms: When nitrogen is available in excess, plants tend to become more succulent, grow taller, and develop heavier heads, which can lead to lodging. Additionally, nitrogen rich, luxuriant plants are more susceptible to insect pests and diseases.

Table 2: Estimated nutrient concentrations in mature leaf tissue of various crop plants

Nutrient	Sufficient or Normal	Deficient	Toxic
N (%)	1-5	-	-
P (%)	0.1-0.4	-	-
K (%)	1-5	-	-
Mg (%)	0.1-0.4	-	-
Ca (%)	0.2-1	-	-
S (%)	0.1-0.4	-	-
Fe (mg/kg)	100-500	<50	>500
Zn (mg/kg)	27-150	10-20	100-400
Mn (mg/kg)	20-300	15-25	300-500
Cu (mg/kg)	5-30	2-5	200-100
Mo (mg/kg)	0.1-2.0	0.03-0.15	>100
B (mg/kg)	10-20	5-30	50-200
Cl (mg/kg)	100-500	<100	500-1000
Ni (mg/kg)	-	<0.1	-

Source: Jones et al. (1993); Tisdale et al. (1997)

Corrective measures: Nitrate (NO_3^-) is particularly susceptible to leaching, especially in sandy soils with high permeability. Therefore, it is recommended to apply nitrogen in split doses. In dryland regions, foliar application of urea at a concentration of 2% is recommended. Nitrogen percentage in different fertilizers are given below:

- Ammonium sulphate: 21.0%
- Ammonium chloride: 25.0%-26.0%
- Ammonium nitrate: 33.0%-34.0%
- Ammonium nitrate sulphate: 30.0%
- Monoammonium phosphate: 11.0%
- Diammonium phosphate: 18.0%-21.0%
- Calcium nitrate: 15.0%
- Urea: 45.0%-46.0%
- Urea ammonium nitrate: 28.0%-32.0%

2. Phosphorus

Phosphorus ranks as the second most prevalent mineral in the human body, accounting for over 20 percent of total body minerals. In plants, phosphorus is primarily taken up by roots as the dihydrogen orthophosphate ion ($H_2PO_4^-$). However, in alkaline to neutral conditions, it is also absorbed as the monohydrogen orthophosphate ion (HPO_4^{2-}). Additionally, phosphorus is absorbed in organic forms, such as phytins and nucleic acids. In phosphorus sufficient plants, the phosphorus content typically ranges from 0.1 to 0.4 percent by weight, which is about one-fifth to one-tenth of the nitrogen or potassium content.

Functions: Once phosphorus enters a plant, it is integrated into essential compounds such as nucleic acids (RNA and DNA), phosphoproteins, phospholipids, sugar phosphates, enzymes, and energy-rich molecules like adenosine triphosphate (ATP) and adenosine diphosphate (ADP). ATP and ADP serve as the plant's energy currency, with ATP storing approximately 12 kcal of energy per mole in two pyrophosphate bonds between three phosphorus atoms. Key processes that rely on ATP include the membrane generation electrical potentials, respiration, and synthesis of cellulose, hemicellulose, pectin, lignin, lipids, phospholipids, proteins, and nucleic acids. Phosphorus thus plays a crucial role in energy transfer, photosynthesis, sugar and starch transformation, nutrient transport within the plant, and the transmission of genetic traits across generations.

Deficiency Symptoms: Plants with phosphorus levels below 0.1 percent or 1000 ppm are typically considered phosphorus deficient. Due to phosphorus' high mobility within the plant, it readily moves from old tissues to meristematic areas, causing older leaves to show deficiency symptoms first. Phosphorus deficiency is often linked to the accumulation of carbohydrates, leading to dark green leaves. Severe phosphorus deficiency significantly restricts the growth of both plant tops and roots. Affected plants tend to be thin, upright, and spindly, with sparse foliage. Lateral bud development is suppressed, and leaves become narrow, forming sharp angles with the stem. The foliage may take on a bluish-green hue, and if the deficiency persists, older leaves may turn bronze or develop reddish purple tips and margins.

Toxicity Symptoms: Excessive phosphorus can lead to deficiencies in trace elements, particularly iron (Fe) and zinc (Zn).

Corrective measures: Phosphorus is typically applied as a basal application using band placement. Common phosphatic fertilizers include single super phosphate (SSP): 16–22% P_2O_5; rock phosphate: 30–38% P_2O_5, triple super phosphate: 46% P_2O_5, diammonium phosphate: 46% P_2O_5, monoammonium phosphate: 52% P_2O_5, and basic slag: 10–20% P_2O_5.

3. Potassium

Potassium is absorbed by plant roots in the form of potassium ions (K^+). The concentration of phosphorus in healthy plant tissues typically ranges between 1-5 percent, which is comparable to nitrogen levels. A distinct characteristic of potassium is that plants can absorb it in large quantities without showing signs of toxicity, a phenomenon known as luxury consumption. Additionally, potassium serves a crucial regulatory function in plant metabolism and growth, though it is not a structural element of the plant itself.

Functions

- Potassium controls the opening and closing of stomata, which are vital for photosynthesis, nutrient and water movement, and regulating plant temperature.
- It is essential for facilitating the transport of water and nutrients via the xylem. When potassium levels are low, the transport of nitrate, phosphate, calcium, magnesium, and amino acids is disrupted.
- Potassium promotes root growth and enhances drought resistance.
- It helps neutralize organic anions and various compounds within the plant, maintaining cytoplasmic pH between 7 and 8, which is ideal for many enzymatic reactions.
- It activates over 60 enzymes by exposing active sites, including starch synthetase, which is responsible for starch synthesis. High potassium levels allow efficient starch movement from production to storage sites. Potassium is also involved in activating the protein forming enzyme nitrate reductase.
- Potassium improves crop quality, enhancing the physical characteristics and shelf life of fruits and vegetables.
- It helps reduce crop lodging and increases winter hardiness.
- Potassium also boosts plant's resistance to diseases.

Deficiency Symptoms: Potassium deficiency symptoms initially appear on older leaves. The leaf margins begin to show chlorosis, followed by browning and scorching at the tips, which gradually spreads inward. In potassium deficient plants, growth is slow and stunted, and they are more prone to lodging. Seeds and fruits from potassium-deficient plants are often shrivelled.

Corrective measures: Potassium chloride, also known as murate of potash, makes up 99% of potassium fertilizer use, with potassium sulphate accounting for the remaining 1%. Murate of potash contains about 60% K_2O (50% K), while sulphate of potash provides 50% K_2O (41.5% K). Certain crops, including fruit trees, tobacco, grapes, cotton, sugarcane, tomatoes, potatoes, strawberries, onions, and cucumbers, are sensitive to high chloride levels. In contrast, crops like oil palm and coconut tend to thrive with higher chloride concentrations. Potassium fertilizers are typically applied as a basal dose; however, split applications are recommended for sandy soils. In Andhra Pradesh, split applications of nitrogen and potassium are advised for lighter soils. Potassium nitrate, containing 44% K_2O (37% K) and 13% nitrogen, is often recommended for foliar application on horticultural crops and fruit trees. Additionally, in winter (rabi season), farmers in Andhra Pradesh apply potassium nitrate sprays to improve cold tolerance.

4. Calcium

Calcium is a secondary nutrient taken up by plant roots in the form of calcium ions (Ca^{2+}). Its concentration typically ranges from 0.2 to 1.0 percent in plants with sufficient calcium levels. Although calcium is one of the most abundant plant available nutrients in soils, its deficiency is more frequently observed in acidic tropical soils.

Functions

- Calcium is a component of calcium pectate in the cell wall, helping to maintain membrane integrity. In seeds, it exists as calcium phytate.
- It is crucial for meristems growth and proper functioning of root tips.
- Calcium protects root cells from ion imbalances, low pH, and toxic ions such as aluminium.
- It neutralizes the charges of harmful acidic molecules like citric acid, phosphoric acid, malic acid, and oxalic acid, which can hinder plant growth.
- Calcium is essential in cell division (mitosis) and aids in maintaining structure of chromosome.
- It activates enzymes such as phospholipase, arginine kinase, amylase, and adenosine triphosphatase (ATPase).

Deficiency Symptoms: Plants with a calcium content below 0.1 percent (or 1000 ppm) are typically considered calcium deficient. However, calcium deficiency is rarely observed on a large scale due to its abundance in soil. Since calcium is immobile within plants and cannot be easily move from older to younger leaves, deficiency symptoms primarily appear in the younger tissues. Under calcium deficiency, root growth is stunted, and young leaves may appear distorted, small, and abnormally green. Affected leaves often become cup-shaped and crinkled, while terminal buds deteriorate, leading to petiole breakdown. Other signs include desiccation of terminal buds and weakened stem structure. Common calcium deficiency symptoms include blossom end rot in tomatoes and bitter pit in apples.

Corrective measures: In calcium deficient acidic soils, calcium plays a crucial role as a plant nutrient. Applying calcium carbonate or calcium sulphate at a rate of 2 to 4 quintals per hectare in furrows can boost crop yields by up to 48%.

5. Magnesium

Magnesium is absorbed by plants in the form of magnesium ions (Mg^{2+}). In plants with sufficient magnesium levels, their concentration typically ranges from 0.1 to 0.4 percent. A significant portion of magnesium is bound with organic anions, such as malate. Magnesium is also a key component of chlorophyll, where it forms a complex as Mg-porphyrin, with a single magnesium atom bound to four pyrrole rings.

Functions

Magnesium, as an essential component of chlorophyll, plays a critical role in photosynthesis.

$$6CO_2 + 6H_2O \xrightarrow{\text{Chlorophyll, sunlight, phosphate energy}} C_6H_{12}O_6 + 6O_2$$

- It also serves as an activator for numerous enzyme systems involved in metabolism of carbohydrate and nucleic acids synthesis.
- Magnesium facilitates the uptake and transport of phosphorus, as well as the movement of sugars within the plant.

Deficiency Symptoms: Magnesium is mobile within plants, with deficiency symptoms typically appearing first on older leaves. Plants suffering from magnesium deficiency usually contain less than 0.1 percent magnesium. Common symptoms include chlorosis and streaky or patchy discoloration on older leaves. As the deficiency progresses, affected leaves may become smaller and curl upwards along the edges. In severe cases, the tissue in these leaves may dry out and die. In some vegetable crops, interveinal chlorosis may be accompanied by shades of red, orange, or purple. Additionally, "grass tetany", a nutritional disorder, is often observed in cattle grazing on magnesium deficient pastures that have been heavily fertilized with potassium.

Corrective measures: Use dolomitic limes stone [$CaMg(CO_3)_2$], magnesia MgO (55% Mg), basic slag (3-4% Mg) at a dose of 30-50 kg/ha.

6. Sulphur

Sulphur is absorbed by plants primarily in the form of sulphate ions (SO_4^{2-}), though some studies suggest that sulphur can also be absorbed through the foliage as sulphur dioxide (SO_2). Similar to phosphorus and magnesium, sulphur concentration in healthy plants typically falls between 0.1 and 0.4 percent. Due to widespread sulphur deficiencies observed in Indian soils, it is often considered the fourth major nutrient after nitrogen, phosphorus, and potassium.

Functions

- Sulphur is a vital component of sulphur-containing amino acids such as cysteine, cystine, and methionine.
- It is a key constituent of ferredoxin-containing nitrogenase, which is involved in Biological Nitrogen Fixation (BNF) and other electron transfer processes.

- Sulphur participates in the metabolic functions of vitamins like biotin, thiamine, and coenzyme A.
- It contributes to the synthesis of glucosides in mustard oil.
- Sulphur is also vital for enhancing oil quality in oilseed crops.

Deficiency Symptoms: Depending on the crop, plants with a sulphur content of less than 0.1 to 0.21 percent are considered sulphur-deficient. Additionally, crop plants with a N:S exceeding 16:1 is also likely to be deficient in sulphur. Since sulphur is immobile within the plant, deficiency symptoms first appear on younger growth. The most noticeable symptom is the loss of the typical green colour in young leaves, which is then followed by chlorosis. In Brassicas, which are particularly sensitive to sulphur deficiency, leaf lamina growth is restricted, leaf margins curl, leading to cupping, and growth points are arrested. Older leaves become puckered, with raised areas between the veins, and may develop orange or reddish hues before prematurely shedding.

Corrective measures

- Apply elemental sulphur or gypsum, especially in alkaline soils.
- Use sulphur-containing fertilizers, such as single super phosphate (12-14% S), magnesium sulphate (30% S), or ammonium sulphate (24.2% S).
- For addressing sulphur deficiencies in standing crops, apply sulphate-based salts as foliar sprays, like ferrous sulphate (32.8% S) or ferrous ammonium sulphate (16% S).

Micronutrients

1. Iron

Iron is absorbed in plants as ferrous ions (Fe^{2+}). A concentration of 100-500 mg/kg in mature leaf tissues is considered adequate for optimal crop growth. As a transition metal, iron exists in two oxidation states, Fe (II) and Fe (III), within plants and forms complexes with organic ligands. Its variable valency makes it essential in biological redox processes.

Functions

- Iron is a component of two protein groups: (a) Heme proteins, which contain an Fe-porphyrin complex as a prosthetic group, such as cytochrome oxidase, catalase, peroxidase, and leghemoglobin; and (b) Fe-S proteins, where iron is bound to the thiol group of cysteine or inorganic sulfur, as in ferredoxin.
- Iron activates several enzymes, including coproporphyrinogen oxidase and aminolevolinic acid synthetase.
- It is essential for metabolism of nucleic acids.
- Iron is essential for maintenance and synthesis of chlorophyll in plants.

Deficiency Symptoms: Plants with less than 50 ppm of iron are typically considered iron-deficient. Research has shown that the Fe^{2+} content, rather than the total iron content, is a more accurate indicator of iron deficiency, with 30 ppm of Fe^{2+} being the critical threshold. Iron deficiency leads to interveinal chlorosis, which initially appears on younger leaves, while the leaf margins and veins remain green. In cases of severe deficiency, plant growth stops, and the entire plant may become necrotic.

2. Manganese

Manganese is absorbrd by plants in the form of manganous ions (Mn^{2+}). Healthy mature plants with sufficient manganese levels contain between 20 and 300 ppm of Mn. As a transition metal, manganese primarily exists in the Mn (II) form within plants but can easily be oxidized to Mn (III) and Mn (IV) forms. Due to its variable redox state, manganese plays a crucial role in photosynthesis and the detoxification of superoxide free radicals.

Functions

- Manganese is an essential component of the water-splitting enzyme linked with photosystem II. As a result, manganese deficiency can negatively impact photosynthesis and oxygen evolution.
- It is a key part of superoxide dismutase (Mn-SOD), which plays a vital role in protecting cells from the harmful effects of superoxide free radicals. Mn-SOD is found in mitochondria, peroxisomes, and glyoxysomes.
- Manganese is also involved in the tricarboxylic acid (TCA) cycle, participating in both oxidative and non-oxidative decarboxylation reactions, with the malic enzyme catalyzing the reaction in conjunction with nicotinamide adenine dinucleotide (NAD) as:

$$\text{Malate} + \text{NADP}^+ \longrightarrow \text{Pyruvate} + \text{NADPH} + \text{H}^+ + \text{CO}_2$$

and isocitrate dehydrogenase catalyzing the reaction:

$$\text{Isocitrate} + \text{NADP}^+ \longrightarrow \text{Oxalosuccinate} + \text{NADPH}$$

Deficiency symptoms: Manganese deficient plants typically have less than 25 ppm of manganese. The deficiency symptoms are more pronounced on the middle leaves than on younger ones, as manganese is preferentially translocated to the younger tissues. In dicotyledonous plants, interveinal chlorosis is marked by the emergence of chlorotic and necrotic spots in the areas between the veins. In monocotyledonous plants such as cereals, manganese deficiency manifests as greenish-grey spots, flecks, and stripes on the lower leaves, known as "grey speck." Chlorotic areas of the leaves eventually become necrotic, turning red, reddish-brown, or brown. Symptoms of manganese deficiency are commonly referred to as grey speck in oats, speckled yellow in sugar beets, marsh spot in peas, pahala blight in sugarcane, and frenching of tung trees.

3. *Zinc*

Plants take up zinc in the form of zinc ions (Zn^{2+}). Mature plants with adequate zinc levels typically contain between 27 and 150 ppm of Zn. Because zinc does not exhibit variable valency, it does not directly influence redox processes.

Table 3: Zinc-containing enzymes and their biological functions

Zinc-containing enzyme	Biological function
Carbonic anhydrase (CA)	Zinc, located in the cytoplasm and chloroplasts, aids in transfer of CO_2/HCO_3^- for carbon fixation during photosynthesis. Severe zinc deficiency can inhibit photosynthesis, and carbonic anhydrase is often used as a diagnostic tool for identifying zinc deficiency.
Alcoholic dehydrogenase	Zinc also plays a crucial role in anaerobic root respiration, catalyzing the conversion of acetaldehyde to ethanol. In waterlogged conditions, anaerobic root metabolism is compromised when zinc is deficient.
Superoxide dismutase (SOD)	Zinc-copper superoxide dismutase (Zn-Cu-SOD) detoxifies superoxide radicals, protecting the lipids and proteins in membranes from oxidation. A reduction in SOD activity due to zinc deficiency exacerbates oxidative processes, such as lipid oxidation and the degradation of indole-3-acetic acid (IAA).

Functions

- Zinc is a key component of three enzymes, which are detailed along with their biological functions in Table 3.
- It is associated in the synthesis of indole-3-acetic acid, gibberellic acid metabolism, and RNA synthesis.
- Due to its strong affinity for sulfhydryl groups, zinc plays a crucial role in stabilizing and orienting the structure of membrane proteins.
- Additionally, zinc affects the translocation and transport of phosphorus in plants. When zinc is deficient, excessive phosphorus translocation can occur, leading to phosphorus toxicity.

Deficiency Symptoms: Plants with zinc levels below 15 ppm are considered zinc-deficient. Typical deficiency symptoms include interveinal chlorosis that initially appears on young leaves, a decrease in the size of these leaves, which may cluster closely together, and bronzing. Additionally, the foliage may exhibit purple, violet, reddish-brown, or brown discoloration. In dicotyledonous plants, zinc deficiency symptoms include short internodes (known as rosetting) and reduced leaf expansion (referred to as little leaf). Various names have been assigned to symptoms associated with zinc deficiency, such as Khaira disease in rice, which is reported in India (also known as Hadda disease in Pakistan, Akagare type II in Japan, and Taya-taya and Apaya-pula in the Philippines). Other names include white bud in maize, mottle leaf (little leaf) or frenching in citrus, and little leaf in cotton. Fields affected by Khaira disease in rice exhibit a coppery-brown appearance and are characterized by absent or missing hills. The initial signs of zinc deficiency emerge in seedlings that are 3 to 4 weeks old, characterized by the development of reddish-brown pigmentation on the young leaves. This pigmentation begins in the centre of the leaf and gradually intensifies, eventually spreading across the entire leaf surface. The affected tissues become papery and necrotic, and in cases of severe deficiency, the entire plant may collapse, leading to stunted growth.

4. *Copper*

Similar to other micronutrient cations, copper is taken up by plant roots in the form of cupric ions (Cu^{2+}). In plants with adequate copper levels, concentrations range from 5 to 30 ppm, while toxicity can arise when copper concentrations are between 20 and 100 ppm (Table 2). Copper is a transition element found in plants as a constituent of various proteins and enzymes in the Cu (II) form, and it can alter its valency based on equilibrium conditions.

$$Cu^{2+} + e^{-} \longrightarrow Cu^{+}$$

Functions: Copper is an essential component of several enzymes, which are detailed along with their biochemical functions in Table 4.

- Copper is vital for enhancing disease resistance in plants.
- It also contributes to the fertility of male flowers.

Table 4: Copper-containing enzymes and their biochemical role

Copper-containing enzyme	Biochemical role
Plastocyanin	Copper is found in chloroplasts, where it forms a part of the electron transport system in photosystem II, thereby playing a vital role in photosynthesis.
Superoxide dismutase (Cu-Zn-SOD)	In the cytoplasm, mitochondria, and chloroplasts, it aids in detoxifying superoxide radicals produced during photosynthesis. Within glyoxysomes, Cu-Zn superoxide dismutase (Cu-Zn-SOD) is believed to help regulate lipid peroxidation, which contributes to the process of senescence.
Diamine oxidase	Found in the apoplasts of the epidermis and xylem tissues, copper serves as a transport system for hydrogen peroxide (H_2O_2) in peroxidase activity during lignification.
Polyphenol oxidase	Copper is responsible for lignin biosynthesis, and a decrease in polyphenol oxidase activity due to deficiency of copper can lead to delayed flowering and late maturity.
Ascorbate oxidase	Found in cell walls and the cytoplasm, copper catalyzes the oxidation of ascorbic acid to dehydroascorbic acid, with the Cu^{2+} component of the enzyme reducing molecular oxygen to water. A reduction in ascorbate oxidase activity linked to lower copper availability serves as a sensitive indicator for diagnosing copper deficiency.

Deficiency Symptoms: Plants with copper levels below 5 ppm are classified as copper-deficient. The primary consequences of copper deficiency include sterility in male flowers, delayed flowering, and premature senescence. Characteristic symptoms of copper deficiency include chlorosis in young shoot tissues, white tips, reclamation disease, necrosis, leaf distortion, and dieback. Necrosis of the apical meristems leads to shoot elongation in cereals and the development of auxiliary shoots in dicotyledonous plants.

5. *Molybdenum*

Molybdenum is the only heavy transition metal absorbed by plants in the form of molybdate ions (MoO_4^{2-}). A healthy plant with sufficient molybdenum contains between 0.1 and 2 ppm of this nutrient. In oxidative environments within the plant system, molybdenum typically exists as Mo (VI) and can be reduced to Mo (V) and Mo (IV). The ability of molybdenum to exist in various valence states gives molybdenum a significant biochemical role.

Functions: Molybdenum serves as a key component of several enzymes, including nitrate reductase, nitrogenase, xanthine oxidase/dehydrogenase, and sulphite oxidase. Due to these enzymes, molybdenum plays several important roles in plant metabolism which are as follows:

- Biological nitrogen fixation (BNF) is facilitated by the molybdenum-containing enzyme nitrogenase, which is composed of a Mo-Fe-S protein and an Fe-S cluster protein. This enzyme transfers electrons directly to nitrogen (N_2). Due to its role in BNF, molybdenum is particularly essential for nodulated legumes.
- In the cytoplasm, nitrate is reduced by the enzyme nitrate reductase (NR) through the transfer of electrons from molybdenum to nitrate (NO_3^-). The close relationship between molybdenum supply, nitrate reductase activity (NRA), and plant growth makes NRA a reliable indicator of molybdenum status in plants.
- Molybdenum plays a vital role in protein biosynthesis by influencing the activity of ribonuclease and alanine aminotransferase.
- It also affects the viability and formation of pollen as well as anthers development.

Deficiency Symptoms: Molybdenum deficiency in plants is typically identified when Mo concentrations fall below 0.1 ppm. The symptoms are often similar to nitrogen deficiency. In plants with reticulate venation, early signs of Mo deficiency include chlorotic mottling between the veins on older or middle leaves. In *Brassica* species, symptoms become evident in 3 to 4 week old plants. Mo-deficient cauliflower displays chlorotic mottling and cupping of the middle leaves, with severe cases leading to scorching and withering, starting at the leaf margins and progressing across the lamina, leaving only the petioles. This condition in cauliflower is known as "whip-tail."

6. *Boron*

Boron is primarily absorbed by plants as boric acid (H_3BO_3), although it can also be taken up in its anionic forms such as dihydrogen borate ($H_2BO_3^-$), monohydrogen borate (HBO_3^{2-}), and borate (BO_3^{3-}) under high pH conditions. In boron-sufficient plants, the B-content ranges from 10 to 200 ppm. Unlike other elements, boron does not serve as a component of enzymes nor does it activate them. One of the key characteristics of boron, particularly boric acid, is its ability to form stable complexes with organic compounds that have a cisdiol configuration.

Functions

- Boron is crucial for formation and stabilization of cell wall, lignification, and differentiation of xylem. As a result, boron deficiency leads to alterations in the chemical composition and ultrastructure of the cell wall, toxic phenols accumulation, inhibition of synthesis of lignin, and reduction in indole acetic acid (IAA) production. This decrease in IAA can trigger calcium deficiency.
- Boron enhances drought tolerance in crops, with regular boric acid applications helping mitigate the negative impacts of drought.
- It is essential for pollen germination and pollen tube growth.
- Boron aids ion uptake by increasing the activity of plasma-membrane-bound H+ ATPase (H^+ adenosine triphosphatase).
- It also promotes potassium transport in guard cells and regulates opening of stomata.

Deficiency Symptoms: Plants with boron concentrations ranging from 5 to 30 ppm may be considered as deficient in boron. The critical range of deficiency for boron ranges from 5 to 10 ppm in grasses and 20 to 70 ppm in most dicotyledonous plants. Boron deficiency symptoms typically appear in terminal buds or the younger leaves; which may become discoloured and, in severe cases, die. Shortened internodes result in a bushy or rosette-like appearance, while the thickening of stems and petioles can lead to the characteristic cracked stem in celery. Common boron deficiency in crops include:

- Heart rot of sugarbeet and mangold
- Browning or hollow stem in cauliflower
- Top sickness in tobacco
- Internal cork in apples.

7. *Nickel*

Plants absorbed nickel in the form of nickel ions (Ni^{2+}). In nickel-sufficient plants, its concentration ranges from 0.1 to 10 ppm. While nickel typically exists in the Ni (II) state, it can also occur in Ni (I) and Ni (III) states.

Functions

- Nickel plays a key role in nitrogen metabolism by regulating urease activity. In cases where urea is used as the only nitrogen fertilizer for foliar application and nickel availability is low, reduced urease activity can result in urea toxicity, leading to severe necrosis in root tips.
- In free-living Rhizobia, sufficient nickel enhances hydrogenase activity, ensuring efficient nutrient transport to seeds or grains.

Deficiency Symptoms: The critical threshold for nickel deficiency in barley shoots is 0.1 ppm, with lower levels causing reduced dry matter production, lower amino acid content, and nitrate accumulation.

8. *Chlorine*

Chlorine is abundant in nature and is absorbed by plants in the form of chloride ions (Cl^-). In healthy plants, the typical chloride concentration ranges from 100 to 500 ppm. Due to its widespread availability and supplementation through various fertilizers, such as murate of potash, chlorine is often overlooked. Additionally, no reports of chlorine deficiency have been documented in India.

Functions

- Chlorine regulates osmoregulation, contributing to elongation of cell and opening of stomata, and is essential for charge balance in higher plants.
- It serves as a cofactor in the manganese-containing water-splitting enzyme of photosystem II.
- Adequate chlorine levels help suppress various plant diseases, such as grey leaf spot in coconut palms, take-all and common root rot in wheat, *Fusarium* root rot and common root rot in barley, stalk rot in corn, stem rot and sheath blight in rice, hollow heart and brown center in potatoes, *Fusarium* yellows in celery, and downy mildew in millet.
- Chlorine improves the nutritional quality of vegetables by reducing nitrate concentration in plant tissues.

Deficiency Symptoms: Plants with chlorine levels below 100 ppm are typically considered deficient. The deficiency symptoms of chlorine, which include leaf wilting, curling of leaflets, and chlorosis, closely resemble those seen in manganese deficiency.

Deficiency identification

A concise guide to deficiency symptoms as outlined by Finck and Geer (1992) is presented in Table 5. However, accurately interpreting these symptoms demands considerable field experience and should be supported by both soil and plant analyses. Additionally, it is important to recognize the influence of other factors that may cause symptoms resembling those of different nutrient deficiencies.

Table 5: Nutrient deficiency symptoms

Nutrient	Deficiency symptoms
	Symptoms first appears on older leaves
Nitrogen	Chlorosis starting from leaf tips
Phosphorus	Reddish colour on stems or green leaves
Potassium	Necrosis on leaf margins
Magnesium	Chlorosis mainly between veins (which remain green)
Manganese	Brownish, greyish, whitish spots (e.g. on cereals)
	Symptoms first appears on younger leaves
Sulphur	Mottled yellow green leaves with yellowish veins
Iron	Mottled yellow green leaves with green veins
Manganese	Brownish black spots (e.g. on legumes, potatoes)
Copper	Youngest leaf has white tip
Boron	Youngest leaf is brownish or dead (e.g. on beet)

Source: Finck (1992)

Beneficial elements

1. *Sodium*

Sodium is predominantly stored in vacuoles, where it contributes to maintaining the cell's solute potential. In certain plants, such as sugar beet, and turnip, sodium can substitute for some essential potassium functions. Additionally, it helps enhance the water balance in plants that experience limited water availability or grow in arid climates.

2. *Silicon*

Silicon plays a positive role in the growth of rice and sugarcane by reinforcing and strengthening cell walls. It also increases the physiological availability of zinc in plants, helping to mitigate phosphorus toxicity associated with zinc deficiency. Furthermore, silicon promotes the growth and yield of cereal crops through indirect effects, such as reducing vulnerability to lodging, lowering the incidence of infections caused by root parasites and pathogens, and preventing iron and manganese toxicity.

3. *Cobalt*

Cobalt boosts nitrogen fixation in legumes and enhances the nutritional quality of forage crops for ruminants. It serves as a metal component of the coenzyme cobalamin, which is essential for the activity of methionine synthetase, ribonucleotide reductase, and methylmalonyl-coenzyme A mutase in *Rhizobium* and *Bradyrhizobium* species.

4. *Selenium*

The chemistry of selenium (Se) resembles that of sulphur (S), with selenium existing in Se^{2-}, Se^{4+}, and Se^{6+} states. Sulphate and selenate compete for the same uptake sites in plant roots, so an increased supply of sulphate can significantly diminish selenium absorption. While selenium is essential for animals, it is not for plants. To prevent selenium deficiency in animals, dietary concentrations should range from 0.1 to 0.3 mg/kg of dry matter. In livestock, selenium deficiency can lead to conditions such as white muscle disease or nutritional muscular dystrophy.

5. *Aluminium*

In plant species that exhibit high tolerance to aluminium (Al) and possess a strong capacity for its uptake, the positive effects of Al on growth stem from its ability to mitigate the toxicity of phosphorus, zinc, and copper.

6. *Vanadium*

Vanadium (V) promotes nitrogen fixation in leguminous plants, exhibiting chemical properties similar to those of molybdenum (Mo). It is believed that vanadium may partially take over the functions of Mo within the nitrogenase system of Azotobacter.

Nutrient Movement in Soil

Transpiration in plants is responsible for the movement of water from the soil into the plant. As water is continuously absorbed by the roots, it creates a hydraulic gradient that facilitates the transport of dissolved nutrients from the soil solution to the plant roots. However, the varying demands for nutrients and water by plants can lead to either an accumulation or depletion of nutrients near the root surface, establishing concentration gradients that either draw nutrients toward or away from the roots. There are two distinct processes involved in transporting nutrients to plant roots. The movement of soluble nutrients along with the soil solution can occur over significantly greater distances in a given time frame than the movement of nutrients through the soil solution itself. This movement of nutrients with water is referred to as "mass flow," while the movement of nutrients through water is called "diffusion." Additionally, as plant roots grow, they penetrate the soil and displace it, coming into direct contact with the soil and facilitating nutrient uptake through contact exchange. Therefore, three recognized mechanisms for nutrient movement are (i) mass flow, (ii) diffusion, and (iii) root interception, each with varying degrees of significance as outlined in Table 6.

Table 6: Relative importance of root interception, mass flow, and diffusion in meeting the nutrient needs of corn grown in a fertile Alfisol silt loam

Nutrient	Amount needed for 9500 kg grain/ha	Approximate amount supplied by		
		Root interception	Mass flow	Diffusion
Nitrogen	190	2	150	38
Phosphorus	40	1	2	37
Potassium	195	4	35	156
Calcium	40	60	150	0
Magnesium	45	15	100	0
Sulphur	22	1	65	0

Source: Barber (1995)

1. *Mass Flow*

Mass flow refers to the transport of nutrients through the soil towards plant roots via the movement of water, driven by the plant's uptake of water. The rate of nutrient transport by mass flow (denoted as J) is determined by the water flux into the root (v) and the concentration of the nutrient in the soil solution (C), represented as $J = v \times C$.

Typical nutrient concentrations in the soil solution of agricultural soils are approximately 10^{-2} moles per litre for calcium (Ca), between 10^{-3} to 10^{-4} moles per litre for potassium (K), and between 10^{-5} to 10^{-6} moles per litre for phosphorus (P). For a crop yielding 10 tonnes of dry matter per hectare and using 300 mL of water per gram of dry matter produced, this water volume would transport roughly 1200 kg of calcium, 12 to 120 kg of potassium, and 0.9 to 9 kg of phosphorus. The calculation for calcium serves as an example for these estimations.

Total dry matter produced = 10 t/ha = 10×10^3 kg/ha = 10^4 kg/ha = $10^4\times10^3$ g/ha = 10^7 g/ha

Amount of water used for producing one gram of dry matter = 300 ml

Amount of water used for producing 10^7 g of dry matter = $10^7\times300$ ml = 3×10^9 ml = 3×10^6 L

Concentration of Ca in soil solution = 10^{-2} M/L = $10^{-2}\times40$ g/L = 4×10^{-1} g/L

Amount of Ca moved along with water to plant root = 3×10^6 L$\times4\times10^{-1}$ g/L = 12×10^5 g = 12×10^2 kg = 1200 kg

The amount of calcium transported by mass flow is approximately 1200 kg/ha. The contributions of mass flow for other nutrients can be calculated in a similar manner.

In reality, crops typically absorb and transport about 10 to 30 kg of calcium per hectare, 50 to 250 kg of potassium per hectare, and 20 to 30 kg of phosphorus per hectare. This highlights the significant role of mass flow in delivering calcium to plant roots. Generally, mass flow supplies an excess of calcium and magnesium to plants in most soils, and the same applies to other mobile nutrients such as nitrate (NO_3^-) and sulfate (SO_4^{2-}). However, the relative importance of mass flow depends on the nutrient concentration in the soil solution. Any nutrient not delivered via mass flow is supplemented through diffusion or root interception.

2. *Diffusion*

When mass flow is unable to provide adequate amounts of a nutrient, and uptake by the plant continues, the nutrient concentration at the root surface decreases, creating a concentration gradient. In response to this gradient, ions move from areas of higher concentration to areas of lower concentration through random thermal motion, known as Brownian movement. The movement of nutrient ions due to this concentration gradient is called diffusion. The rate of nutrient ion diffusion across a unit area in a given time (F), in the presence of a concentration gradient (dC/dx), is described by Fick's first law of diffusion for steady-state conditions.

$F = -D.(dC/dx)$ where, D is the diffusion coefficient.

The equation is applicable to a linear gradient, with the negative sign indicating that the movement is directed from an area of higher concentration to area of lower concentration.

In the case of transient diffusion, the change in concentration over time at a fixed distance x is described by Fick's second law.

$$\delta C/\delta t = D(\delta^2 C/\delta x^2)$$

This equation applies to a linear diffusion.

In the case of cylindrical plant roots, the equation is applied as following equation:

$$\delta C/\delta t = (1/r).(\delta/\delta r).[r.D.(\delta C/\delta r)]$$

where, r is the radial distance from the centre to the plant root to its surface.

3. *Root Interception*

The term "root interception," introduced by Stanley A. Barber from Purdue University, USA, refers to the process where soil nutrients are made available for absorption as roots grow into the soil, rather than requiring the nutrients to move towards the root surface. The amount of nutrients supplied through root interception is considered to be equivalent to the quantity present in the volume of soil corresponding to the root's volume. Typically, the root volume of many crop species accounts for less than 1% of the soil volume in the 0-20 cm soil depth, so calculations are based on the nutrient availability within 1% of the soil volume. As the root system expands, more soil, along with the soil solution and surfaces holding adsorbed ions, becomes accessible to the roots. The adsorption of these ions occurs through a contact exchange mechanism, as described by Jenny et al., (1939). In nutrient-poor soils, root interception can be enhanced by mycorrhiza, a symbiotic relationship between certain fungi and plant roots.

References

Arnon, D.I. and Stout, P.R. (1939) The Essentiality of Certain Elements in Minute Quantity for Plants with Special Reference to Copper. Plant Physiology, 14, 371-375.

Barber, Nick. "What constitutes good prescribing?." Bmj 310.6984 (1995): 923-925.

Bray, R.H. and Kurtz, L.T. (1945) Determination of Total Organic and Available Forms of Phosphorus in Soils. Soil Science, 59, 39-45. http://dx.doi.org/10.1097/00010694-194501000-00006.

Finck, Robert, and Lars Erik De Geer. "Field gamma spectrometry and soil sampling of Chernobyl fallout at reference areas in Russia." High resolution field gamma spectrometry and its application to problems in environmental radiology. Department of Radiation Physics, Malmö and Lund, Lund University (1992): 115-123.

http://dx.doi.org/10.1104/pp.14.2.371

Jenny, Hans, Roy Overstreet, and A. D. Ayers. "Contact depletion of barley roots as revealed by radioactive indicators." Soil science 48.1 (1939): 9-24.

Jones, J.W., 1993. Decision support systems for agricultural development. In: Penning de Vries, F.W.T., Teng, P., Metselaar, K. (Eds.), Systems Approaches for Agricultural Development. Kluwer Academic Press, Boston, pp. 459–471.

Nicholas DJD (1961) Minor mineral nutrients. Annu Rev Plant Physiol 12:63–90. https://doi.org/10.1146/annurev.pp.12.060161.000431.

Tisdale, S. L., Nelson, W. L., Beaton, J. D. and Havlin, J. L. (1993). Soil Fertility and Fertilizers, 5th edn. New York, NY: Macmillan.

3

Role of Essential Plant Nutrients

Nitrogen

1. Nitrogen is a vital component of proteins and chlorophyll and is found in numerous compounds critical to plant metabolism, such as nucleotides, phosphatides, alkaloids, enzymes, hormones, and vitamins. It is, therefore, a fundamental element for life.
2. Enhances the dark green color of plants.
3. Stimulates the growth of leaves, stems, and other vegetative parts, though it leads to a relatively smaller root system.
4. Promotes rapid early development.
5. Enhances the quality and succulence of leafy vegetables and fodder crops.
6. Increases the protein content in food and fodder crops.
7. Plays a significant role in the efficient utilization of potassium, phosphorus, and other essential nutrients.

Phosphorus

1. Phosphorus is a key component of nucleic acids, phytin, and phospholipids. The adequate early supply of phosphorus in the life of plant is crucial for reproductive structures development .
2. It is also a vital part of many enzymes involved in energy transfer, carbohydrate metabolism, fat metabolism, and respiration (carbohydrate catabolism) in plants. Phosphorus plays a vital role in cell division and growth.
3. Promotes early root development and growth, aiding in the quick establishment of seedlings.
4. Ensures a rapid and vigorous start for plants, strengthens stems, and reduces the risk of lodging.
5. Accelerates crop maturity, particularly in cereals, and counteracts the effects of excessive nitrogen.
6. Stimulates flowering and supports formation of seed.
7. Improves the grain-to-straw ratio.
8. Enhances the food grains quality and other crops.

9. In legumes, phosphorus boosts rhizobia activity, increasing the formation of root nodules and enabling greater atmospheric nitrogen fixation. Without sufficient phosphorus, legumes may also experience deficiencies in nitrogen and potassium.
10. In some cases, excessive phosphorus can lead to trace element deficiencies, particularly of iron and zinc, but it may also mitigate the negative effects of over-liming.

Potassium

Unlike other major nutrients, potassium does not form part of essential plant compounds such as proteins, chlorophyll, fats, or carbohydrates that are involved in plant metabolism. As a result, its role is less straightforward to define. Potassium exists in solution within the cell sap and, due to its solubility, can be easily removed from plant tissues with water. The essential functions of potassium are as follows:

1. Increases plant vigor and enhances resistance to diseases.
2. In cereals, such as rice and wheat, potassium strengthens stems and reduces lodging.
3. Improves winter hardiness in legumes and other crops.
4. Regulates water balance within plant cells and reduces water loss by maintaining equilibrium between anabolism, respiration, and transpiration. This helps minimize wilting and improves water use efficiency.
5. Essential for the synthesis and movement of starches and sugars, making it particularly important for crops like potatoes, sweet potatoes, turnips, bananas, and tapioca.
6. Aids in the formation of proteins and chlorophyll.
7. Increases grain and seed plumpness.
8. Acts as a catalyst for enzyme activity, such as that of diastase.
9. Mitigates the harmful effects of excessive nitrogen in plants, making a balanced nitrogen-to-potassium ratio crucial for optimal plant nutrition.
10. Enhances the quality of final products, improving the quality of tobacco leaves, fiber crops, and the flavor, size, and shelf life of fruits. However, excessive potassium in citrus fruits may negatively impact their quality.

Calcium

1. Calcium is a key component of the cell wall, contributing to the stiffness of plant stems.
2. Promotes early root growth and development.
3. Serves as a base for neutralizing organic acids, often considered toxic byproducts within the plant.
4. Essential for activating growth points, particularly at root tips. However, calcium does not move easily from older to younger plant parts, which is

why deficiency symptoms typically appear first at the tips of shoots and roots.

5. Influences water regulation in plants, affects the protein-to-carbohydrate ratio in fat metabolism, and plays a role in many other physiological processes. The effects of calcium often oppose those of potassium, making a balanced potassium-to-calcium ratio important for maintaining proper water balance.
6. Enhances the uptake of other nutrients, especially nitrogen and trace elements like iron, boron, zinc, copper, and manganese, by improving soil pH.
7. Promotes seed production.

Magnesium

1. As a component of chlorophyll, magnesium is vital for all green plants and helps maintain the dark green color of leaves.
2. It is involved in the synthesis of proteins, carbohydrates, fats, and vitamins, and plays a role in certain catalytic reactions within enzyme systems.
3. Functions as a carrier of phosphorus within plants, particularly aiding in the formation of seeds with high oil content, thereby promoting the production of oils and fats.
4. Assists in the translocation of starches.
5. Regulates the absorption of other nutrients and maintains the plant's base nutrient balance.

Sulphur

1. Although sulphur is not a component of chlorophyll, it plays a role in its formation and supports vegetative growth in plants.
2. Sulphur is an essential part of many proteins, enzymes, and certain volatile compounds like mustard oil. It also contributes to the redox system in respiration and the activation of enzymes.
3. Enhances root growth.
4. Stimulates seed development.
5. Promotes the formation of nodules on the roots of legumes.

Boron

1. Boron's primary function is associated with calcium metabolism, aiding both its uptake by roots and its efficient utilization within the plant.
2. Helps maintain calcium in a soluble form and enhances its mobility within the plant.
3. Regulates the potassium-to-calcium ratio in the plant.
4. Assists in nitrogen absorption.

5. Boron is a component of cell membranes and is essential for cell division. In the absence of boron, cell division ceases at growing points, although individual cells may continue to grow. Thus, boron deficiency symptoms, similar to calcium deficiency, are most noticeable at the tips of shoots and roots.
6. Plays a role in precipitating excess cations, buffering, maintaining conducting tissues, and regulating the effects of other elements.
7. Necessary for the translocation of sugars in plants.
8. Supports the root's vascular system by promoting the formation of rootlets, which supply nodule bacteria with carbohydrate nutrients, preventing them from becoming parasitic.
9. Influences a variety of other plant functions, including active salt absorption, hormone transport, flowering and fruiting processes, pollen germination, carbohydrate and nitrogen metabolism, pectin metabolism, respiration, and water regulation.

Manganese

1. Manganese plays a role closely linked to that of iron and aids in the movement of iron within the plant.
2. Contributes to chlorophyll formation.
3. Serves as a catalyst in oxidation-reduction reactions in plant tissues, and as part of enzymes, it supports respiration and protein synthesis in chloroplasts.
4. An adequate manganese supply can sometimes offset the negative effects of poor soil aeration.

Iron

1. Although iron is not a part of chlorophyll, it plays a crucial role in its formation. Iron deficiency leads to chlorosis, or yellowing between leaf veins, and because iron is immobile in plants, the deficiency first appears in younger leaves at the growing tips.
2. Aids in the absorption of other nutrients.
3. As a component of enzyme systems involved in oxidation-reduction reactions, iron regulates key processes such as respiration, photosynthesis, and the reduction of nitrates and sulphates—essential for plant growth and reproduction.
4. Necessary for the synthesis of proteins within chloroplasts.

Zinc

1. Zinc is a component of various enzyme systems that control metabolic processes in plants, such as oxidation-reduction reactions and chlorophyll formation.
2. Plays a role in the synthesis of certain growth hormones in plants.

3. Supports the reproductive processes of specific plants.
4. Involved in water uptake and overall water management within the plant.

Molybdenum

1. Functions in enzyme systems that facilitate oxidation-reduction reactions, particularly the conversion of nitrates to ammonia, which is a precursor for amino acid and protein synthesis in plant cells.
2. Essential for both symbiotic and non-symbiotic atmospheric nitrogen fixation, enhancing the ability of legumes to efficiently fix nitrogen from the atmosphere.

Copper

1. Copper forms various complexes with amino acids and proteins in plants.
2. It serves as an "electron carrier" in enzymes that facilitate oxidation-reduction reactions within the plant.
3. Aids in the utilization of iron during chlorophyll synthesis.
4. When applied in larger amounts, it can neutralize harmful conditions in certain peat soils, such as those found in Kerala, by precipitating or inactivating specific toxins present in the soil.

Chlorine

Chlorine was identified as an essential plant nutrient in 1954 by T. C. Broyer and his colleagues at the University of California (USA). The necessity of chlorine for optimal plant growth has been established in crops such as sugar beets, carrots, lettuce, barley, wheat, cotton, and clovers. However, the specific role of chlorine in plant nutrition has not yet been fully elucidated. Regarding soil fertility, plants require approximately 1 kg of chlorine for every 4,000 kg of dry matter produced. For high yielding crops like sugarcane and sugar beets, this can translate to a need for 5.6 kg or more of chlorine per hectare.

4

Methods to Identify Deficiency Symptoms Using Plants

Visual method or hunger signs of plants

The method for diagnosing a plant's fertilizer needs is by observing visible "hunger signs." These signs are based on the fact that plants experiencing severe nutrient deficiencies or excesses often develop distinct and characteristic symptoms, particularly in the leaves. Nutrient deficiencies usually result in specific abnormal leaf colors. Although these signs are relatively easy to spot, identifying the specific nutrient deficiency in natural field conditions can be challenging, requiring experience and field practice.

Nutrient deficiency symptoms can appear first either in the lower or upper parts of the plant, depending on the mobility of the nutrient. Based on their mobility, major plant nutrients are classified as follows:

(i) *Nutrients that show deficiency symptoms in older or lower leaves (highly mobile nutrients)*

Nitrogen, Phosphorus, Magnesium, Potassium, and Zinc

(ii) *Nutrients that exhibit deficiency symptoms in younger or bud leaves (less mobile nutrients)*

Calcium, Boron, Copper, Manganese, Sulphur, and Iron

The common deficiency symptoms exhibited by essential plant nutrients are summarized as follows:

1. *Nitrogen*
   - Leaves turn yellow or pale green.
   - Lower leaves dry out or experience "firing."
   - Plants exhibit stunted growth.
2. *Phosphrous*
   - Leaves turn red or purple.
   - Growth slows, and plants mature late.
   - Lower leaves dry out, with purple discoloration between the veins.
   - Leaf petioles display a purple hue.

3. *Potassium*
    - Margins and tips of lower leaves appear scorched or burned.
    - Leaves thicken and curl.
    - Deficiency initially appears in the wetter areas of the field.
4. *Calcium*
    - Young leaves at the terminal bud become wrinkled and die back at the tips and edges.
    - The stalk eventually dies at the terminal bud.
5. *Magnesium*
    - Overall yellowing of leaves begins with the lower leaves and later moves upward.
    - Leaf veins remain green.
    - In cotton, lower leaves develop a purplish-red color between the green veins.
6. *Sulphur*
    - Young leaves show veins and tissues of light green color.
    - Growth is slow.
7. *Iron*
    - Young leaves appear chlorotic, with the main veins remaining green.
    - Stalks are short and slender.
    - Young growing tissues may die back, resulting in a rosetting appearance.
8. *Manganese*
    - Dead tissue appears as spots scattered across young leaves.
    - The smallest veins typically remain green, resulting in a checkered or reticulated appearance.
9. *Zinc*
    - Leaves display yellow streaks between the veins.
    - Lower leaves die off first.
    - The plant is severely stunted.
10. *Boron*
    - Young leaves at the terminal bud turn light green at the base and eventually break down.
    - As growth continues, leaves may twist, and the stalks ultimately die at the terminal bud.
    - In cauliflower, browning of the curd and lesions in the pith occurs.

11. *Copper*
    - Young leaves exhibit permanent wilting without spots or significant chlorosis.
    - Normally green leaves may turn yellow and show signs of chlorosis.
    - In grains, conditions known as white tip or yellow tip disease occur, leading to dwarfed or distorted heads.
12. *Molybdenum*
    - Lower leaves in tomato plants display mottling, necrosis, and curling at the edges.
    - In oats, leaves bend backward and later break at the affected areas, showing necrosis.
13. *Chlorine*
    - Leaves show symptoms such as wilting at the tips of the leaf blades, chlorosis, bronzing, and necrosis in areas close to the wilting.

5

Nitrogen

Occurrence, distribution and content

The primary source of nitrogen utilized by plants is the inert gas nitrogen, which makes up approximately 78% (by volume) of the Earth's atmosphere. Among all essential nutrients, nitrogen has been extensively studied and continues to attract significant attention. In the ploughed layer of most cultivated soils, nitrogen content typically ranges from 0.02% to 0.04%. Indian soils tend to have low nitrogen levels due to limited organic matter accumulation, result of the tropical climate. Soils formed in cool, humid climates generally exhibit higher nitrogen content compared to those developed in arid conditions. On average, black soils contain more nitrogen (0.08%) than red soils (0.03%).

Factors affecting nitrogen content of in soils

1. *Climate*
   - Temperature: Colder temperatures generally lead to higher nitrogen content, as they promote greater organic matter accumulation and slower decomposition rates.
   - Water supply: Nitrogen content in the soil increases with water availability up to field capacity.
2. *Cultivation effect:* Nitrogen content tends to decrease with cultivation due to the loss of organic matter through decomposition.
3. *Soil texture*: Soils with fine textures typically have high nitrogen content.
4. *Soil profile depth*: Surface soils generally contain more nitrogen than deeper soils.

Forms of nitrogen in soil

Soil nitrogen content varies widely, ranging from below 0.02% in subsoils to over 2.5% in peat soils. The nitrogen found in soil can be categorized into inorganic (approximately 2%) and organic forms (around 98%).

1. *Inorganic compounds of nitrogen*: The inorganic nitrogen forms present in soil include ammonium (NH_4^+), nitrate (NO_3^-), nitrite (NO_2^-), nitric oxide (NO), nitrous oxide (N_2O), and elemental nitrogen. Ammonium, nitrate, and nitrite are crucial for soil fertility and account for 2% to 5% of the total nitrogen content.

2. *Organic compounds of nitrogen*: These are found as free amino acids, consolidated amino acids or proteins, amino sugars, and other unidentified materials. These include compounds formed through the reaction of NH_4^+ with lignin, the polymerization of nitrogen compounds and quinones, and the condensation of amines and sugars. Bound amino acids typically represent 20% to 40%, amino sugars (hexosamines) about 5% to 10%, and purine and pyrimidine derivatives make up 1% or less.

Nitrogen transformations in soil

Plants primarily absorb nitrogen in the forms of NH_4^+ and NO_3^-. Nitrate is the predominant source due to its higher concentration compared to ammonium, and it is readily available for root uptake. Different plants have varying preferences: potatoes, sugar beets, and pineapples can utilize both forms; tomatoes, celery, and bush beans favor NO_3^-, while rice and blueberries prefer NH_4^+. The uptake of NO_3^- -N is generally high and is enhanced under low pH conditions. In contrast, NH_4^+-N experiences fewer losses due to leaching and denitrification, and its uptake is optimal at neutral pH levels. However, when plants are provided with NH_4^+-N, it can lead to increased soil acidity.

Mineralization of N compounds

Nitrogen mineralization refers to the process of converting organic nitrogen into mineral forms such as NH_4^+, NO_3^-, and NO_2^-. When organic residues with a carbon-to-nitrogen (C:N) ratio greater than 30 are added to the soil, nitrogen immobilization occurs. Conversely, if the CN ratio is narrow, specifically less than 20 (as seen with legume residues), mineralization takes place. This process generally occurs in three stages:

1. *Aminisation*: Heterotrophic soil microbes, predominantly bacteria like *Pseudomonas* and *Bacillus*, play a key role in breaking down proteins in neutral and alkaline soils. In acidic conditions, fungi become more prevalent. During this step, proteins undergo hydrolytic decomposition, resulting in the release of amines and amino acids.

   $$\text{Proteins} \longrightarrow R\text{-}NH_2 + CO_2 + \text{Energy} + \text{Other products}$$

2. *Ammonification:* The released amines and amino acids are subsequently utilized by different groups of heterotrophic organisms, resulting in the production of ammoniacal compounds. This process is known as ammonification.

   $$R\text{-}NH_2 + HOH \longrightarrow NH_3 + R\text{-}OH + \text{Energy}$$

   $$NH_3 + H_2O \longrightarrow NH_4^+ + OH^-$$

The ammonium generated through this process can be nitrified into nitrites and nitrates, which are then available for plant uptake.

Behaviour and transformation of released ammonium in the soil

1. Nitrification can convert ammonia into nitrites and nitrates.
2. Higher plants directly absorbed ammonia.
3. Heterotrophic organisms utilized ammonia for further decomposition of residues of organic carbon.
4. It might be fixed in a biologically unavailable form within the structure of certain expanding clay minerals.
5. Volatilization release ammonia back into the atmosphere.

Fixation of ammonia

In agricultural soils, fixed ammonium ions account for 5% to 20% of the total nitrogen, with an average of around 10%. When fertilizers that contain or produce ammonium are applied to the soil, the majority of the ammonium becomes adsorbed onto cation exchange sites. However, some of it is prone to strong adsorption within the interlayer spaces of certain 2:1 layer silicate clay minerals that have an expanding lattice structure. The sequence of ammonium fixation is as follows: smectites < fine-grained mica < vermiculite. The ionic diameter of ammonium ion is 2.96 Å, which is similar to the 2.8 Å diameter of interlayer spaces in 2:1 type clay minerals. Fixed ammonium is not readily available to plants and microorganisms; it can only be released by exchanging NH_4^+ for interlayer cations in the expanded lattices of clay minerals. Cations that can expand the lattice, such as Ca^{2+}, Mg^{2+}, and Na^+, can replace the fixed ammonium.

The presence of K^+ (with an ionic diameter of 2.68 Å) often limits the fixation of NH_4^+, as K^+ can also occupy these fixation sites. Therefore, it has been proposed that applying K fertilizers before application of NH_4^+ is an effective strategy to minimize fixation of NH_4^+. Additionally, the organic matter can chemically fixed or bound ammonium, with these reactions occurring more frequently in the presence of oxygen and at elevated soil pH levels. In organic soils that possess a high capacity for fixation, this reaction could lead to significant losses of available nitrogen, suggesting the need for fertilizers other than those providing NH_4^+. Nevertheless, the fixed NH_4^+ will gradually be released through mineralization.

Factors influencing fixation of ammonium

1. *Amount and type of clay:* The fixation of NH_4^+ improves with the clay content, especially with 2:1 type mineral such as vermiculite, fine-grained mica, and smectite.
2. *Soil moisture content:* The fixation of NH_4^+ is affected by temperature and moisture content of soil. Processes like freezing and thawing enhance fixation, while alternating wetting and drying cycles, as well as freezing and thawing, are thought to promote the stability of recently fixed NH_4^+.

3. Amount of K^+: The presence of K^+ limits NH_4^+ fixation, as K^+ competes for the same fixation sites.
4. *Soil depth*: NH_4^+ fixation is typically greater in subsoils compared to surface soils, attributed to high clay content and low rates of nitrification in deeper layers.
5. *Organic matter content*: Increased organic matter content correlates with greater NH_4^+ fixation.
6. *Nitrifying bacteria population*: Nitrifying bacteria convert NH_4^+ to nitrite and nitrate, thereby reducing the likelihood of NH_4^+ fixation.

Nitrification

The oxidation of NH_4^+ which is released during ammonification to nitrate is known as nitrification. This process is facilitated by nitrifying bacteria, commonly referred to as nitrifiers. Nitrification occurs in two steps: first, NH_4^+ is converted to nitrite (NO_2^-), and then nitrite is transformed into nitrate (NO_3^-). The conversion of NH_4^+ to nitrite is primarily carried out by a group of obligate autotrophic bacteria called Nitrosomonas, as represented by the following reaction:

$$2NH_4^+ + 3O_2 \longrightarrow 2NO_2^- + 2H_2O + 4H^+$$

The conversion from nitrite to nitrate is performed by another type of bacteria, known as Nitrobacter, as follows:

$$2NO_2^- + O_2 \longrightarrow 2NO_3^-$$

Factors influencing nitrification

1. *Availability of ammonium ions*: Since ammonium ions serve as the substrate for nitrifying bacteria, an adequate supply of these ions is essential for the nitrification process.
2. *Population of nitrifying bacteria*: The rate of nitrification is significantly impacted by the number of nitrifying organisms, given similar conditions of temperature, moisture, and ammonium availability.
3. *Soil pH*: Nitrification occurs optimally within a pH range of 5.5 to 10.0, with the ideal level being around 8.5.
4. *Soil aeration*: Nitrifying bacteria are aerobic, meaning they require a sufficient oxygen supply to thrive.
5. *Moisture content of soil*: The rate of nitrification is greatly influenced by soil moisture levels, which are inhibited by both excessively low and high moisture content.
6. *Temperature*: Nitrification is minimal near freezing temperatures and increases rapidly up to 35°C.
7. *CN ratio*: When organic residues with a wide CN ratio are introduced, general decomposer organisms dominate, making nitrifiers less active.

8. *Pesticides*: Certain pesticides can negatively affect nitrifying bacteria, as these organisms are sensitive to some chemical treatments.

Transformation of nitrate nitrogen

Nitrate, whether introduced through commercial fertilizers or generated via the nitrification of ammonium, can follow several pathways:

1. It may be taken up by plants.
2. It can be utilized by soil microorganisms.
3. Due to its high mobility, it may be lost through leaching in the presence of excess water.
4. It can be lost from the soil in gaseous form through a process known as denitrification, where nitrate is reduced.

Nitrate nitrogen losses through leaching

Nitrogen loss from the soil through leaching represents a loss from the soil-plant system, reducing the efficiency of applied nitrogen. This loss can range from 2% to 50% of the nitrogen applied, with over 90% typically occurring as nitrate (NO_3^-) due to the limited anion exchange capacity of soils compared to their cation exchange capacity. As a result, nitrate movement is closely tied to water movement in the soil. The extent of nitrogen leaching is influenced by factors such as texture of soil, hydrological conditions, depth of water table, permeability of soil, and hydraulic conductivity. Strategies to minimize leaching losses include improving water retention through the addition of organic matter, applying nitrogen in split doses, using controlled irrigation, and incorporating nitrification inhibitors like N-serve, or slow-release nitrogen fertilizers such as neem-coated urea or sulphur-coated urea.

Denitrification

Denitrification is a process that occurs in oxygen-deprived (anoxic) soils, where bacteria reduce nitrate (NO_3^-) and nitrite (NO_2^-), resulting in the release of gases such as nitric oxide (NO), nitrous oxide (N_2O), and nitrogen (N_2). When soils become waterlogged and oxygen is unavailable, anaerobic decomposition begins. Certain anaerobic bacteria can utilize nitrates and nitrites as a source of oxygen, leading to the production of nitrous oxide and nitrogen gases. The biochemical pathway responsible for these gaseous losses can be described as follows:

$$2NO_3^- \xrightarrow[-2(O)]{+4H^+} 2NO_2^- \xrightarrow[-2(O)]{} 2NO \xrightarrow[-2(O)]{} N_2O \xrightarrow[-O]{} N_2$$

Facultative anaerobic bacteria, such as *Pseudomonas*, *Bacillus denitrificans*, and *Paracoccus* are primarily responsible for denitrification, along with autotrophic bacteria like *Thiobacillus denitrificans*.

Nitrification inhibitors

Several compounds have been developed to slow the release and/or nitrification of applied nitrogen, aligning nitrogen availability with crop demand, and reducing losses through leaching or denitrification. These inhibitors limit the activity of *Nitrosomonas*, keeping nitrogen in the ammonium (NH_4^+) form.

Natural form of nitrification inhibitors: They are neem cake (*Azadirachta indica*), karanj cake (*Pongamia glabra*), neem oil and nimin.

Chemical form of nitrification inhibitors: N-Serve (nitrapyrin: 2-chloro-6-trichloromethyl pyridine), Dicyandiamide (DCD), AM (2-amino-4-chloromethyl pyrimidine), ATC (4-amino-1,2,4 triazole), sulphathiazole, thiourea, hydroquinone, calcium carbide (CaC_2), and phenyl phosphorodiamidate (PPD).

However, the widespread use of these chemicals in India is limited due to their high costs.

Immobilization of nitrogen

Immobilization is the reverse process of mineralization. It occurs when large amounts of crop residues with a high carbon-to-nitrogen ratio (C:N > 30:1), such as coconut coir or cereal straw, decompose in the soil. The high carbohydrate content in these residues causes a rapid increase in the microbial population. As microbes form new cells, they utilize nitrogen and other essential nutrients from the soil to build protoplasm. This results in a temporary reduction of inorganic nitrogen available for crops. To prevent nitrogen deficiency, additional nitrogen fertilizer can be applied to compensate for immobilization and meet crop needs. The immobilized nitrogen is eventually released back into the soil as the organic matter undergoes mineralization.

Fixation of nitrogen

Bacteria can convert atmospheric nitrogen into ammonia through an enzyme called nitrogenase, a process known as biological nitrogen fixation (BNF). This contributes approximately 140-170 million tonnes of nitrogen annually, while fertilizers provide only 65 million tonnes per year.

Types of Nitrogen-Fixing Organisms

There are three primary groups of nitrogen-fixing organisms

1. *Symbiotic Nitrogen Fixers*

These bacteria fix nitrogen in association with plants, forming specialized structures.

a) *Legume-Rhizobium Symbiosis*: Certain plants in the leguminosae family form symbiotic relationships with bacteria from the genus *Rhizobium* that fix atmospheric nitrogen. The bacteria infect the plant roots, moving through an infection thread into the root cortex, forming tiny outgrowth called root nodules. In species like *Sesbania rostrata*, nodules can also form on the stem. This relationship is highly host-specific, meaning a specific *Rhizobium* species associates with particular legume plants. Inside the nodules, the bacteria transform into bacteroids, which use nitrogenase enzymes embedded in leghaemoglobin to fix nitrogen. This symbiosis is mutually beneficial, with the bacteria providing nitrogen and the plant supplying nutrients. The host specificity of different *Rhizobium* species, or groups of legumes nodulated by a single *Rhizobium* species, is known as a "cross-inoculation group."

b) Symbiosis with Non-Legumes

- Non-Legume - *Frankia* Symbiosis: The actinomycete genus *Frankia* establishes symbiotic relationships with forest trees from the *Casuarinaceae* and *Myricaceae* families.

c) Azolla: Anabaena Symbiosis: Azolla is a freshwater fern that floats on the surface of water. All species of Azolla contain an algal symbiont known as *Anabaena azollae* within a specialized cavity on the upper surface of their leaves. In paddy fields, this symbiosis can fix nitrogen up to 30-40 kg N ha^{-1}.

2. *Non-Symbiotic Nitrogen Fixers*

Rhizobacteria, such as *Azospirillum*, *Acetobacter*, *Azotobacter*, *Flavobacterium*, and *Pseudomonas* can utilize root exudates as their energy source.

3. *Free-Living Nitrogen Fixers*

Numerous soil bacteria can convert molecular nitrogen (N_2) to ammonium (NH_4^+) independently, without forming associations with plants. These bacteria contribute approximately 10-15 kg of nitrogen per hectare and also act as plant growth-promoting rhizobacteria (PGPR).

$$N_2 + 6H^+ + 6e^- \xrightarrow{\text{Nitrogenase}} 2NH_3$$

$$N_2 + 8H^+ + 8e^- \xrightarrow{\text{Nitrogenase}} 2NH_3 + H_2$$

The primary product of the nitrogenase-catalyzed reaction is ammonia (NH_3), with intermediate products as follows:

$$\underset{\text{Nitrogen}}{N_2} \xrightarrow{+2H^+} \underset{\text{Diimide}}{[NH{=}NH]} \xrightarrow{+2H^+} \underset{\text{Hydrazine}}{[NH_2{=}NH_2]} \xrightarrow{+2H^+} \underset{\text{Ammonia}}{2\,NH_3}$$

The requirements for biological nitrogen fixation include:

i) An active nitrogenase enzyme system
ii) Magnesium ions (Mg^{2+}) and a continuous supply of ATP
iii) A suitable electron donor, or strong reducing agent
iv) Low levels of oxygen

Nitrogen balance sheet

1. *Soil nitrogen gains*: Soil can acquire nitrogen through four primary pathways:
   a) *Nitrogen fixation by symbiotic bacteria*: The bacteria that form symbiotic relationships with host plants, particularly legumes, fix nitrogen within root nodules. These bacteria, primarily from the genus *Rhizobium*, obtain the necessary energy from the carbohydrates produced by the host plant.
   b) *Azofication*: Heterotrophic, free-living bacteria such as *Azotobacter* and *Clostridium* contribute to nitrogen fixation through a process known as azofication. These microorganisms obtain their energy from decomposing organic matter.
   c) *Rainfall contributions*: It is estimated that, on average, 5 kg of nitrogen per hectare is added to the soil annually through precipitation.
   d) *Additions from manures and fertilizers*: A significant portion of nitrogen gain in soil comes from green manures, crop residues, organic manures, and nitrogen fertilizers.
2. *Losses of nitrogen from soil*: Nitrogen can be removed from the soil through several mechanisms:
   (i) *Removal by crops and microorganisms*: While this may not be considered a true loss, it represents a reduction of nitrogen available in the soil. Crop uptake is a significant pathway for nitrogen loss.
   (ii) *Leaching*: Nitrogen can be lost via leaching, primarily in the forms of nitrate (NO_3^-) and ammonium (NH_4^+). Over 90% of these losses occur as nitrate due to the soil's lower anion exchange capacity (AEC) compared to cation exchange capacity (CEC). Soil texture, moisture levels, depth of water table, permeability, and hydraulic conductivity of the soil influence the extent of leaching losses.
   (iii) *Gaseous nitrogen losses*: Nitrogen can be lost as ammonia (NH_3) through volatilization or as nitrogen oxides (NO), nitrous oxide (N_2O), and dinitrogen (N_2) via denitrification.
   (iv) *Volatilization of ammonia*: Nitrogen losses through ammonia can range from 5% to 35%, with lower losses occurring in flooded soils. An increase in soil pH towards alkaline conditions enhances volatilization.

(v) *Denitrification*: Due to saturation, when the redox potential in the soil falls below +250 mV, microbes like *Thiobacillus denitrificans* use oxygen from nitrogen compounds, leading to losses of 10% to 40%.

(vi) *Chemodenitrification:* This term refers to non-microbial processes that reduce nitrogen in the soil to gaseous forms.

(vii) *Immobilization:* This soil microorganisms converts inorganic form of nitrogen into organic forms of nitrogen.

(viii) *Ammonium fixation by clay minerals*: Ammonium (NH_4^+) is strongly adsorbed in the interlayer spaces of 2:1 clay minerals.

(ix) *Soil erosion and runoff*: Runoff can lead to nitrogen losses as high as 70%.

To assess nitrogen balance, one can calculate it by considering both nitrogen additions and losses:

Nitrogen Balance = (Amount of fertilizer nitrogen added + nitrogen added through crop residues, irrigation, rainwater, and manure) – (Crop uptake of nitrogen + nitrogen losses through various processes + nitrogen remaining in the soil).

6

Phosphorus

Phosphorus (P), a crucial nutrient for all living organisms, is an essential component of the molecular structures forming genes and chromosomes. In soils, phosphorus primarily exists as sedimentary formations and unprocessed ore, commonly referred to as phosphate rock. Major reserves of phosphate rock are found in countries such as China, Morocco, Jordan, Egypt, Western Sahara, South Africa, and the USA. In India, notable reserves include 13.5 million tons of high-grade phosphate rock (30% P_2O_5) located in Jhamarkotra, Rajasthan.

In Indian soils, total phosphorus content generally varies between 100 and 200 ppm, with higher levels typically present in soils derived from granite and gneiss, followed by those from shale. Total phosphorus in soils includes both inorganic and organic forms.

Forms of phosphorus in soil

1. *Inorganic Phosphorus*: In soils, inorganic phosphorus is generally more abundant than organic phosphorus (with the exception of organic soils), making up approximately 54 to 84% of the total phosphorus content. This inorganic phosphorus can be divided into various pools.

Soil solution phosphorus: Plants absorb phosphorus mainly in the form of primary and secondary orthophosphate ions ($H_2PO_4^-$, HPO_4^{2-}), with a specific form depending on the soil's pH level. The concentration of phosphorus in the soil solution is very low, typically less than 0.1 ppm.

$$H_2PO_4^- \underset{H^+}{\overset{OH^-}{\longleftrightarrow}} H_2O + HPO_4^{2-} \underset{H^+}{\overset{OH^-}{\longleftrightarrow}} H_2O + PO_4^{3-}$$

Strong acidic solution Strong alkaline solution

$$\text{Soil solution } P^- \longleftrightarrow \text{Labile soil } P^- \longleftrightarrow \text{Non labile P}$$

Readily available phosphorus in the soil solution exists as $H_2PO_4^-$ and HPO_4^{2-} ions, which are released from other phosphorus forms or from phosphorus added to the soil.	⟷	Slowly available (labile) phosphorus includes Fe, Al and Mn phosphates in acidic soils, as well as Ca and Mg phosphates in alkaline soils, which are newly formed.	⟷	Very slowly available phosphorus (non-labile) consists of aged and well-crystallized precipitates of Fe, Al and Mn phosphates, along with stable organic phosphate compounds.

a) *Labile soil phosphorus:* It represents the readily accessible component of the quantity factor and has a high dissociation rate, allowing for the quick replenishment of phosphorus in the soil solution. When labile phosphorus is depleted, non-labile phosphorus can convert to labile form, though this process occurs slowly. Phosphorus associated with aluminium (Al-P), iron (Fe-P), and calcium (Ca-P) forms the main active labile pool of inorganic phosphorus. In soils, Fe-P and Al-P generally account for 1–25% of total phosphorus, while Ca-P can make up 40–50% or even more, especially in neutral to alkaline and calcareous soils.

b) *Non-labile phosphorus*: The less reactive, relatively inactive forms of phosphorus include occluded phosphorus and reductant-soluble phosphorus. The calcium compounds present in soil include Flourapatite [$3Ca_3(PO_4)_2CaF_2$], Carbonate apatite [$3Ca_3(PO_4)_2CaCO_3$], Hydroxy apatite [$3Ca_3(PO_4)_2Ca(OH)_2$], Oxy apatite [$3Ca_3(PO_4)_2CaO$], Tricalcium phosphate [$3Ca_3(PO_4)_2$], Octa calcium phosphate [$Ca_8H_2(PO_4)_5.5H_2O$], Dicalcium phosphate [$CaH\ PO_4.2H_2O$] and Monocalcium phosphate [$Ca(H_2PO_4)_2$].

Among the forms listed above, mono- and dicalcium phosphates are the most readily available for plant uptake. In contrast, apatite minerals are highly insoluble and largely inaccessible to plants.

Phosphorus compounds containing iron and aluminium include Strengite [$FePO_4.2H_2O$], Vivianite [$Fe_3(PO_4)_2.8\ H_2O$], Variscite [$Al\ PO_4.2H_2O$] and Wavellite [$Al_3(OH)_3(PO_4)_2.5H_2O$].

2. *Organic Phosphorus:* In Indian soils, organic phosphorus ranges from 0 to 2160 ppm, accounting for 20–90% of the total soil phosphorus. The main organic phosphorus compounds in soil include:

a) *Inositol phosphates (2–9%):* Phytin, a calcium-magnesium salt of phytic acid (inositol phosphoric acid) with the empirical formula $(CH)_6(H_2PO_4)_6$, constitutes up to 30–40% of soil organic phosphorus.

b) *Phospholipids*: These phosphorus-containing fatty acids make up 1–2% of soil organic phosphorus.

c) *Nucleic acids:* Contribute 1–2% to the soil organic phosphorus pool.

d) Unidentified esters and phosphoproteins

Phosphorus in these forms is mineralized by soil microorganisms, releasing phosphates that plants can readily absorb.

Factors affecting the availability of inorganic phosphorus

Soil pH: Soil pH significantly impacts phosphorus reactions with various ions.

$$H_2PO_4 \underset{H^+}{\overset{OH^-}{\longleftrightarrow}} H_2O + HPO_4^- \underset{H^+}{\overset{OH^-}{\longleftrightarrow}} H_2O + PO_4^{3-}$$

Very acidic solution Very alkaline solution

Plants prefer monovalent forms of phosphorus over divalent or trivalent forms. A soil pH between 6.5 and 7.5 is considered ideal for maximizing phosphorus availability. In highly acidic soils, the concentration of iron and aluminium ions is significantly higher, leading to thc formation of insoluble compounds with $H_2PO_4^-$, which makes phosphorus unavailable. At high pH levels, $H_2PO_4^-$ can react with $CaCO_3$, forming insoluble tricalcium phosphate.

$$\underset{\text{Monocalcium P}}{Ca(H_2PO_4)_2} + CaCO_3 \longrightarrow \underset{\text{Dicalcium P}}{2CaHPO_4} + CO_2 + H_2O$$

$$\underset{\text{Dicalcium P}}{2CaHPO_4} + 2CaCO_3 + H_2O \longrightarrow \underset{\text{Octacalcium P}}{Ca_8H_2(PO_4)_5.5H_2O} + 2CO_2$$

$$\underset{\text{Octacalcium P}}{Ca_8H_2(PO_4)_5.5H_2O} + CaCO_3 \longrightarrow \underset{\text{Dicalcium P}}{3Ca_3(PO_4)_2} + CO_2 + H_2O$$

Phosphorus fixation

Phosphorus fixation is a chemical process that renders phosphorus unavailable to crops, often referred to as "P retention". Phosphate fixation involves the soil removing phosphate from solution, thus reducing its availability for plant root uptake. This phenomenon was first demonstrated by Thomas Way in 1850. Soil pH strongly influences phosphorus fixation. In neutral to alkaline soils (pH 7.0 and above), lime promotes phosphorus precipitation, while in acidic soils (pH below 6.5), phosphorus reacts with iron and aluminum, forming highly insoluble compounds.These processes can be illustrated through chemical equations, with three primary mechanisms involved depending on the soil pH.

1) *In neutral to alkaline soils*

$$\underset{\text{Monocalcium phosphate}}{Ca(H_2PO_4)_2} + 2CaCO_3 \longrightarrow \underset{\text{Tricalcium phosphate}}{Ca_3(PO_4)_2} + 2CO_2 + 2H_2O$$

2) *In acidic soils*

$$Al^{3+} + H_2PO_4^- + 2H_2O \longrightarrow 2H^+ + Al(OH)_2H_2PO_4$$

Dihydrogen P (Soluble) Aluminium Hydoxy P (Insoluble)

As the products of phosphate fixation are largely insoluble, only trace amounts of phosphorus are readily available for plant growth.

3) *In neutral soils*

Phosphorus is fixed within the crystal lattices of 1:1 type clay minerals, such as kaolinite, which have a high anion exchange capacity (AEC) through the anion exchange process.

Factors influencing phosphorus fixation in soil

1) *Clay minerals:* The primary site of phosphorus fixation in soils lies within the clay fraction. The greater the surface area of a specific type of clay, the more likely it is to retain phosphorus.
2) *Clay type*: Phosphorus retention is more pronounced in 1:1 clay compared to 2:1 clay. Soils rich in kaolinitic clay can fix larger amounts of added phosphorus due to their higher anion exchange capacity (AEC).
3) *Iron and aluminium hydrous metal oxides*: Soils containing substantial amounts of iron and aluminium oxides exhibit a higher capacity for phosphorus fixation.
4) *Amorphous colloids*: In Andisols, the presence of significant quantities of amorphous colloids, such as allophanes, which have large negative charges balanced by aluminium cations, leads to phosphorus being fixed to these aluminium ions.
5) *Calcium carbonate*: Calcium carbonate can precipitate phosphates, forming insoluble apatites. Soils with high calcium ion activity tend to have lower phosphorus availability.
6) *Soil pH*: The adsorption of phosphorus by iron and aluminium oxides decreases as pH increases.
7) *Organic matter*: A higher organic matter content reduces phosphorus fixation.
8) *Temperature*: An increase in temperature typically enhances phosphorus fixation.
9) *Over liming*: Excessive liming increases phosphorus fixation by creating more insoluble calcium-phosphate compounds.

Ways for reducing phosphorus fixation

1) *Addition of organic matter*: Incorporating organic residues into the soil affects phosphate reactions and their availability to plants. Organic matter

mitigates the influence of $CaCO_3$ and decreases phosphorus absorption by the soil. It also enhances phosphorus concentration in the soil solution through the mineralization of organic phosphorus and the solubilization of native phosphorus compounds. This effect is particularly pronounced at higher moisture levels. During the decomposition of organic matter, various organic acids are produced that help solubilize phosphates and other phosphorus-bearing minerals, thus reducing phosphorus fixation. This occurs for several reasons:

a) The formation of phospho-humic complexes, which are readily assimilated by plants.
b) The mobilization of phosphate through humate ions via anion exchange.
c) The creation of a protective humus coating on sesquioxide particles, which prevents phosphorus fixation.

2) *Placement of fertilizers*: Fine soil textures lead to greater retention of added phosphorus due to greater reaction speed with the available surface area. When finely ground phosphorus fertilizers are applied to the soil through broadcasting, a larger surface area is exposed, resulting in higher fixation. In contrast, band placement reduces the contact surface between the soil and fertilizer, thereby decreasing the amount of fixation.
3) *Function of Mycorrhiza*: Certain mycorrhizal fungi, known as VAM fungi, colonize plant roots and enhance phosphorus absorption for the plants.
4) *Role of phosphorus solubilizing microorganisms:* Phosphorus solubilizing bacteria such as *Bacillus megatherium* and *Pseudomonas striata*, as well as fungal species like *Aspergillus awamori*, have been shown to increase the availability of fixed phosphorus, potentially adding 30–35 kg of P_2O_5 per hectare under optimal conditions.
5) *Liming of acidic soils*: Approximately one-third of Indian soils are acidic. The beneficial effects of liming these acidic soils stem from the release of fixed phosphorus that is bound to iron and aluminium oxides and hydroxides.

Quantity-Intensity Relationships

Schofield and Taylor (1955) categorized the measurements as follows:

a) *Intensity of supply*: This reflects the concentration of phosphate in the soil solution and is quantified by phosphate potential.
b) *Quantity Factor*: This represents the total phosphorus in the soil, which replenishes the soil solution pool (intensity) as plants absorb phosphorus.

Within a specific range of phosphorus concentrations, where the quantity–intensity (Q/I) relationship is linear, the slope of the curve ($\Delta Q/\Delta I$), known as the phosphate buffering capacity, indicates the soil's resistance to phosphorus depletion. This buffering capacity, or simply "capacity," is generally higher in fine-textured soils than in coarse-textured soils for a given intensity level. Phosphorus availability to plants is determined by the interplay of these factors. As a result, clay soils experience less depletion than sandy soils and can sustain phosphorus availability to crops over extended periods.

References

Schofield, R. K., & Taylor, A. W. (1955). The measurement of soil pH. Soil Science Society of America Journal, 19(2), 164-167.

7

Potassium

Potassium is regarded as one of the three primary nutrients essential for plant growth. In India, 1.33 million tons of potassic fertilizers are consumed, and the gap between potassium removal and its application to crops is observed to be increasing. The average potassium content in the Earth's crust is estimated to be approximately 1.9%, with a range of 0.5 to 2.5%. In tropical soils, the total potassium content is found to be generally low due to the soil's origin, high rainfall, and elevated temperatures.

Potassium sources: Soil potassium originates from the weathering and breakdown of rocks containing potassium-bearing minerals. The primary minerals serving as potassium sources are potassium feldspars, orthoclase and microcline ($KAlSi_3O_8$), muscovite [$KAl_3Si_3O_{10}(OH)_2$], and biotite [$K(MgFe)_3AlSi_3O_{10}(OH)$]. Potassium also exists in soils as part of secondary silicate clays, including illite, vermiculite, and chlorite.

Forms of potassium in soil: Soil potassium is generally classified into four categories according to its availability to crops:

1. Water-Soluble Potassium
2. Readily Exchangeable Potassium
3. Fixed Potassium

Lattice or Inert Reserve Potassium

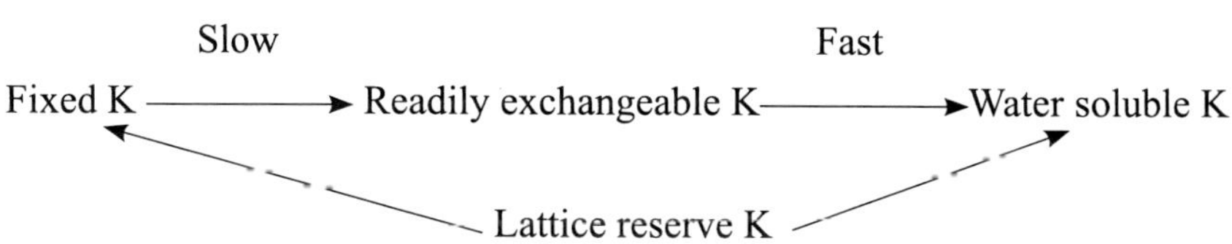

1. *Water-soluble potassium*: Potassium in the soil solution exists as a soluble cation and is in equilibrium with exchangeable potassium
2. *Exchangeable potassium*: This form of potassium is adsorbed onto soil clay particles and can be replaced by neutral salts. In mineral soils, exchangeable potassium typically constitutes less than 1% of the total potassium.
3. *Fixed potassium*: Fixed potassium refers to the portion of added potassium that is strongly held by the soil and is not readily replaced by neutral salts. This form is held between the layers of micaceous minerals.

4. *Lattice Potassium*: Lattice potassium represents the largest portion of total potassium in mineral soils and is found in primary minerals like muscovite, biotite, and potash feldspars, making up about 92–98% of the total potassium content.

Potassium Fixation

The process in which applied or soil solution potassium is converted into non-exchangeable or fixed forms that cannot be extracted with neutral salts is known as potassium fixation. This fixation is most prominent in 2:1 clay, especially those with high levels of illite. It is minimal in soils where kaolinite, chlorite, and unweathered micas are prevalent, moderate in montmorillonite, considerable in illite, and highest in soils dominated by vermiculite. In clay minerals such as weathered mica, vermiculite, and illite three types of adsorption sites can be identified: planar surfaces (p positions), layer edges (e positions), and interlayer spaces (i positions). Potassium binding with organic colloids and kaolinite occurs at the 'p' position, where it is loosely bound and readily exchanged with other cations. Potassium is held at the 'i' position in smectite-rich soils which has the highest specificity for potassium. During dehydration, the lattice layers draw closer, resulting in the loss of water molecules from the adsorbed cations.

The Lattice Hole Theory given by Page and Baver (1940) suggests that the exposed mineral surfaces and the spaces between mineral sheets consist of oxygen ions arranged in a hexagonal pattern. The gap within this hexagon matches the diameter of an oxygen ion (about 2.8 Å). Ions with a similar diameter, like potassium (K^+), which has a radius of approximately 2.66 Å, fit precisely into these lattice holes, where they are tightly bound due to contact with the negative charges within the crystal structure. Similarly, ions such as ammonium (NH_4^+), with a radius close to that of K^+ (around 2.86 Å), are also susceptible to fixation, especially by 2:1 clay minerals.

Factors influencing fixation of potassium

1) *Clay minerals*: Clay minerals primarily influence potassium fixation, depending on their quantity and type. Minerals like illite, weathered mica, and vermiculite contribute to potassium fixation, whereas 1:1 clays like kaolinite generally do not.
2) *Soil reaction*: The fixation capacity of potassium may decrease in the presence of Al^{3+} and aluminium hydroxide cations.
3) *Cation Exchange Capacity (CEC)*: Fine-textured soils, with higher CEC, retain more exchangeable potassium.
4) *Potassium concentration*: In soils with high fixing capacity, higher concentrations of potassium encourage greater fixation.
5) *Wetting and drying*: Drying of soils rich in exchangeable potassium promotes potassium fixation due to air-drying effects.

6) *Freezing and thawing:* This process in moist soils can release fixed potassium.
7) *Subsoil potassium and rooting depth*: The levels of exchangeable potassium vary with soil depth.
8) *Soil temperature*: Higher temperatures facilitate dehydration and contraction of the clay crystal lattice, leading to increased potassium fixation.
9) *Calcium and Magnesium (Exchangeable cations)*: The size of potassium and other replacing ions is significant in potassium fixation. Cations with smaller hydrated ionic radii can more easily enter clay lattices and substitute some of the fixed potassium.

Quantity and Intensity parameters

The intensity, quantity, and buffering capacity concepts proposed by Beckett (1964) provide a valuable framework for assessing a soil's potassium-supplying capacity. The intensity factor measures potassium in the soil solution that is readily accessible for plant uptake. Given that the absorption of potassium is affected by other cations activity such as Ca^{2+} and Mg^{2+} in the soil solution, the potassium activity ratio (AR^{k}) is used instead of just potassium concentration to represent this intensity factor.

$$AR^{K} = \frac{K^{+}}{\sqrt{[Ca^{(2+)}} + \sqrt{Mg^{(2+)}]}}$$

The quantity factor reflects the soil's ability to sustain potassium levels in the soil solution over extended periods or throughout crop growth, primarily due to its exchangeable potassium reserves. The slope of the curve represents the potential buffering capacity, indicating how the soil solution's potassium level (intensity) changes in relation to the amount of labile potassium (quantity).

$$PBC = \frac{\Delta Q}{\Delta I}$$

A wider ratio indicates greater soil buffering. Soils with smectitic clay exhibit higher phosphorus buffering capacity (PBC) compared to those containing kaolinite.

Luxury consumption: It refers to the tendency of certain crops to absorb and accumulate potassium far beyond their actual requirements when available in ample amounts in the soil. This excess uptake, known as luxury consumption, does not significantly enhance crop yields and is often inefficient, particularly noticeable in forage crops.

References

Beckett, P. (1964). Studies on soil potassium II. The 'immediate' Q/I relations of labile potassium in the soil. Journal of Soil Science, 15 (1), 9-23.

Page, J. B., and L. D. Baver. (1940). Ionic size in relation to fixation of cations by colloidal clay. 150-155.

8

Secondary Nutrients

Secondary nutrients, which include magnesium, calcium, and sulphur, often receive less direct attention because they are commonly added unintentionally through fertilizers and irrigation water. For instance, applying 100 kg of ammonium sulphate provides 20 kg of nitrogen and 24 kg of sulphur, while single super phosphate contributes 16 kg of P_2O_5, 21 kg of calcium, and 12 kg of sulphur. Similarly, potassium sulphate supplies 52 kg of K_2O and 18 kg of sulphur. Though required in quantities comparable to primary nutrients, these secondary nutrients are usually applied incidentally along with primary nutrient fertilizers. Lime, containing calcium and magnesium, is often applied to increase soil pH, whereas sulphur compounds are used to decrease it. Irrigation water typically contains around 25 ppm of sulphate (SO_4^{2-}), 50 ppm of calcium, and 5 ppm of magnesium.

I. Calcium

Calcium content and sources in soil: The concentration of calcium in the earth's crust averages around 3.2%, ranging from as low as 0.015% in humid regions to as high as 5% in arid areas, where reduced rainfall and minimal leaching often contribute to higher calcium levels. Primary sources of calcium include weathered materials from rocks and minerals, such as:

- Calcite [$CaCO_3$]: 56.0%
- Apatite [$Ca_{10}(PO_4)_6$]: 50.0%-53.2%
- Dolomite [$CaCO_3.MgCO_3$]: 33.2%
- Gypsum [$CaSO_4.2H_2O$]: 32.2%

Other primary minerals, such as Augite (18.7%), Hornblende (1.6%), Anorthite (10.0–20.0%), and Epidote (22.0–25.0%), also contribute calcium to soils.

Forms of calcium in soil: Calcium is typically the principal cation in well-managed agricultural soils, with its levels generally higher in calcareous soils compared to acidic soils.

1. *Mineral particles*: Calcium is predominantly found in primary minerals such as basic plagioclase (like anorthite) and basic rocks, including basalt and gabbro, all of which release calcium upon weathering.
2. *Calcium carbonate:* The mineral calcite is widely present, along with nodular and amorphous forms of calcium carbonate.

3. *Simple salts*: Calcium also occurs in various simple salts, including calcium chloride, calcium sulphate, calcium nitrate, and calcium bicarbonate.
4. *Exchangeable calcium*: Among the cations in the soil's exchange complex, Ca^{2+} is typically the most abundant.

Fate of released calcium

1. *Uptake by organisms*: Plants like legumes require higher amounts of calcium compared to grasses. Prolonged cropping of calcium-demanding crops, such as alfalfa, particularly on sandy soils low in calcium, can deplete soil calcium levels.
2. *Losses through leaching*: Significant calcium loss occurs as drainage water carries away calcium due to rainfall or irrigation, with losses being greater in light-textured soils compared to heavier soils. Increased rainfall or soil permeability also raises calcium loss, with leaching typically ranging from 84 to 224 kg/ha.
3. *Adsorption by clay colloids*: When calcium levels in soil solution rise suddenly, equilibrium shifts, leading to some calcium being adsorbed by the soil's exchange complex. Conversely, if calcium levels drop due to leaching or crop uptake, adsorbed calcium is released back into the soil solution.
4. *Reprecipitation as secondary calcium compounds*: In arid regions, low rainfall levels promote calcium precipitation as secondary, less-soluble compounds, reducing its availability in soil solution.

Factors influencing the availability of calcium in soils

1. *Total calcium supply*: In highly sandy, acidic soils with low cation exchange capacity, the total calcium supply may be insufficient to meet crop requirements.
2. *Soil pH*: High hydrogen ion (H^+) activity at low soil pH can restrict calcium uptake due to high levels of base unsaturation.
3. *Cation Exchange Capacity (CEC) and Saturation of the Exchange Complex*: The absolute calcium content is less critical than the relative abundance of other cations or the degree of calcium saturation. For instance, montmorillonite requires 70% base saturation to release calcium, while kaolinite requires only 40%.
4. *Type of soil colloid:* The type of clay affects calcium availability, 2:1 clay minerals require a higher degree of saturation for calcium availability compared to 1:1 clays.
5. *Calcium-to-other cations ratio in soil solution*: Calcium availability is influenced by its ratio to other cations in the soil solution. An optimal calcium-to-total cation ratio is 0.10 to 0.15. Maintaining a calcium-to-salt ratio between 0.16 and 0.2 can prevent calcium deficiency disorders like blossom end rot in tomatoes.

II. Magnesium

Plants absorb magnesium from soil solution as the Mg^{2+} ion. Magnesium comprises about 1.93% of the earth's crust, with concentrations varying from 0.1% in coarse sandy soils of humid regions to as high as 4% in fine-textured soils in arid and semi-arid areas. Magnesium in soil originates primarily from the weathering of minerals like biotite, dolomite, hornblende, olivine, epsomite ($MgSO_4.7H_2O$), and serpentine. It is also present in secondary clay minerals such as chlorite, illite, montmorillonite, and vermiculite, and can be especially abundant in soils derived from serpentine bedrock. Magnesium accounts for 12-80% of exchangeable bases in soils, ranking second to calcium, which typically comprises 80-85%. When large amounts of magnesium interact with montmorillonite, this may result in its transformation to vermiculite.

Fate of released magnesium

As primary minerals decompose, magnesium (Mg) is released into the soil, where it may:

a) Leach out with percolating water.
b) Absorbed by living organisms.
c) Adsorbed onto nearby clay colloids,
d) Reprecipitate as secondary minerals in arid areas.

The magnesium available to plants is found in the exchangeable and water-soluble forms. Its availability is influenced by:

1) *Amount of exchangeable Mg*: Soils are considered deficient in magnesium (Mg) when they contain less than 25 to 50 ppm of exchangeable Mg.
2) *Soil pH*: Nutritional issues related to Mg can occur in strongly acidic soils with aluminium (Al^{3+}) saturation levels between 65% and 70%. Additionally, high hydrogen ion (H^+) activity in such soils can negatively impact Mg availability.
3) *Degree of Mg saturation*: A higher degree of Mg saturation correlates with greater availability. However, when exchangeable magnesium exceeds 45%, it can become toxic to plants. Optimal availability occurs when Mg saturation on the exchange complex is between 10% and 20%. Typically, exchangeable magnesium constitutes 4% to 20% of the cation exchange capacity (CEC), and it should not fall below 10%.
4) *Nature of other exchangeable ions*: Magnesium deficiencies in plants can arise in soils with a high calcium (Ca) to magnesium (Mg) ratio, which ideally should not exceed 7:1. Higher levels of exchangeable potassium (K^+) can hinder Mg uptake, with the K/Mg ratio recommended to be below 5:1 for field crops, 3:1 for vegetables and sugar beets, and 2:1 for fruits and greenhouse crops. Ammonium (NH_4^+) also interferes with magnesium nutrition.

5) *Removal by crop*: Leguminous plants tend to deplete more magnesium (Mg) from the soil compared to non-leguminous plants.
6) *Additions through fertilizers, manures, and rain*: Farmyard manure (FYM) and oil cakes are excellent sources of magnesium. Additionally, basic slag contains about 3% to 4% magnesium, while serpentines can provide up to 26% magnesium.

III. Sulphur

Sulphur constitutes about 0.06% of the earth's crust, primarily found as sulphides, sulphates, and in organic compounds with carbon and nitrogen. In temperate regions, over 95% of sulphur in soil is typically organic, while soils in tropical and subtropical regions generally have lower organic matter content. The primary origin of sulphur is sulphide minerals in igneous rocks, with sulphides released through weathering. In humid regions, sulphur is predominantly organic, whereas in arid areas, sulphates of calcium, magnesium, sodium, and potassium are more common. Atmospheric sulphur, released from industrial activities or fuel combustion as sulphur dioxide, also contributes sulphur through rainfall. The frequent application of high-analysis fertilizers, especially sulphur-frcc options like urea, triple super phosphate, ammonium nitrate, and ammonium chloride, can lead to sulphur deficiencies in soils.

Forms of sulphur in soil

Sulphur in soils exists in various forms, both organic and inorganic. The inorganic forms of sulphur include readily soluble sulphates, adsorbed sulphates, insoluble sulphates co-precipitated with calcium carbonate, and reduced inorganic sulphur compounds.

1. *Readily soluble sulphate*: Plants typically absorb sulphur as the sulphate (SO_4^{2-}) ion. The concentration of readily soluble sulphate usually in sulphur deficient soil ranges from 5 to 10 ppm, with sandy soils often containing less than 5 ppm. Sulphate ions associated with monovalent cations like sodium (Na^+) are more susceptible to leaching, while those bound to divalent cations such as magnesium (Mg^{2+}) or calcium (Ca^{2+}) experience moderate losses, and the least losses occur when bound to trivalent cations like aluminium (Al^{3+}) or iron (Fe^{3+}).
2. *Adsorbed Sulphate*: This fraction is significant in soils rich in hydrous oxides of iron and aluminium, such as Oxisols and Ultisols. Adsorbed sulphate can contribute up to one-third of the total sulphur content in these soils.
3. *Sulphate coprecipitated with calcium carbonate*: This form is notably present in calcareous soils.
4. *Sulphides*: Hydrogen sulphide (H_2S) can accumulate in anaerobic conditions of waterlogged soils due to the decomposition of organic matter. Sulphate

(SO_4^{2-}) in the soil acts as an electron acceptor for sulphate-reducing bacteria, which convert it to H_2S. This reduction process occurs at a redox potential lower than -150 mV and within a pH range of 6.5 to 8.5.

5. *Elemental sulphur*: This is an intermediate product formed during the chemical oxidation of sulphides.
6. *Organic form*: In well-drained agricultural soils across humid, temperate, semi-arid, and subtropical regions, the majority of sulphur exists in organic form, accounting for over 90%. The average carbon-to-nitrogen-to-sulphur (C:N:S) ratio is 113:10:1.3 for calcareous soils and 147:10:1.4 for non-calcareous soils. In most soils, the nitrogen-to-sulphur (N:S) ratio typically falls between 6 and 8 and is found in the form of amino acids such as cysteine, cystine, and methionine.

Transformations of sulphur

The transformations of sulphur in soil are diverse and typically cyclical, with organic and inorganic forms being converted through biological processes. When plant and animal residues are added to the soil, they are decomposed by microorganisms, and a portion of sulphur is released as sulphate. However, most of the sulphur is retained in its organic form and is ultimately incorporated into soil humus.

1. *Mineralization*: This process involves the converting organic sulphur to inorganic sulphate (SO_4^{2-}) by microorganisms. The extent of mineralization depends on the sulphur content of the decomposing organic matter and typically occurs at a carbon-to-sulphur (C:S) weight ratio of approximately 200:1. Heterotrophic microorganisms facilitate this process, with the sulphtase enzyme playing a crucial role. Mineralization is generally more pronounced in cultivated soils compared to fallow soils due to the rhizosphere effect.
2. *Immobilization*: Sulphur can become immobilized in soils where the carbon-to-nitrogen-to-sulphur (C:N:S) ratio is excessively wide. When the C:S ratio exceeds 200:1, the immobilization of sulphate (SO_4^{2-}) occurs, particularly when the ratio surpasses 400:1. This typically happens when low-sulphur organic materials are added to soils that are not rich in inorganic sulphur. In such cases, energy rich materials promote microbial growth, leading to the incorporation of inorganic sulphate into microbial biomass. Consequently, inorganic sulphate becomes locked within the microbial tissues, and it only reappears in the soil solution once microbial activity declines.

Oxidation of sulphur

Elemental sulphur, sulphides, and various inorganic sulphur compounds can undergo oxidation in the soil primarily through microbial processes, although some chemical non-biological oxidation occurs to a lesser degree. The biological oxidation rate is influenced by three key factors namely microbial population, characteristics of the sulphur source, and environmental conditions in the soil. The overall reaction can be represented as:

$$CO_2 + S + H_2O_2 + 2H_2O \longrightarrow (CH_2O)n + SO_4^{2-} + 2H^+$$

Two specific classes of bacteria are particularly well-suited for sulphur oxidation namely chemolithotrophic sulphur bacteria, which use the energy released during the oxidation process. Examples include: *Thiobacillus thiooxidans* and *Thiobacillus thioparus*.

Factors influencing oxidation of sulphur in soils

1. *Soil microflora*: An increase in the population of sulphur-oxidizing bacteria, such as *Thiobacillus thiooxidans*, enhances the rate of sulphur oxidation.
2. *Temperature:* The optimal temperature range for this process is between 25°C and 40°C, with sulphur oxidation being sluggish at temperatures below 10°C.
3. *Soil pH*: The ideal pH range for sulphur oxidation is between 6.5 and 7.2.
4. *Soil type and properties*: The oxidation of sulphur is not significantly affected by soil texture or organic matter content. Faster sulphur oxidation is typically observed in well-fertilized soils compared to those low in phosphorus and potassium. *Thiobacilli* require ammonium (NH_4^+) for growth, while nitrate (NO_3^-) can be harmful. Although organic matter is not essential for *Thiobacilli*, as they are autotrophic rather than heterotrophic, it may still play a role in soil health.
5. *Fertilizer interactions:* The application of sulphur in conjunction with nitrogen and phosphorus fertilizers can enhance the rate of sulphur oxidation.

Reduction of sulphur or sulphide injury or H_2S injury

In anaerobic environments, such as waterlogged soils, hydrogen sulphide (H_2S) can accumulate due to the decomposition of organic matter. Additionally, sulphate (SO_4^{2-}) in the soil acts as an electron acceptor for sulphate-reducing bacteria, such as *Desulfovibrio* and *Desulfotomaculum*, leading to the reduction of sulphate to hydrogen sulphide. This sulphate reduction process is influenced by both redox potential and pH levels. Sulphide toxicity tends to occur at a redox potential below -150 mV and within a pH range of 6.5 to 8.5.

In certain rice paddies characterized by high organic matter content and low levels of active metals like iron, free H_2S may be released, posing a risk to rice roots. This harmful effect of H_2S on rice plants in paddies is known as "Akiochi."

Symptoms of Akiochi

- The upper leaves progressively become shorter.
- Lower leaves dry out, turning a dirty yellow with brown spots on the surface.
- Roots appear pale white, mixed with black or decayed roots, giving off a rotten egg smell; when washed and exposed to sunlight, thcy turn reddish.
- Soil pH is around 6.0.
- Bubbles are observed rising from the affected areas.

In normally submerged soils rich in iron (Fe), the hydrogen sulphide (H_2S) produced is largely removed from the solution by reacting with iron to form iron sulphide (FeS_2), which coats the roots and helps prevent decay.

$Fe^{2+} + S^{2-} \longrightarrow FeS_2$ (iron sulphide)

Preventive measures

- Avoid using ammonium sulphate or other sulphur-containing substances.
- Apply iron-rich materials, such as laterite or lake mud, to the soil.
- Improve soil aeration by draining the field.

9

Micronutrients

The term "micronutrients" refers to essential nutrients needed in minimal amounts to develop plants and microorganisms. These micronutrients, also known as "trace elements," include iron, copper, manganese, zinc, boron, molybdenum, nickel, and chlorine. Iron is often present in the soil and plants in higher concentrations than even phosphorus (P) and sulphur (S). Another element, cobalt (Co), is recognized as essential for the growth of certain higher plants, animals, and microorganisms, and it may eventually join the list of essential micronutrients due to its small yet necessary role. Of these nine micronutrients, iron, manganese, zinc, copper, cobalt, and nickel typically exist as cations in soil, while boron, molybdenum, and chlorine exist as anions. Therefore, the chemical behaviours of micronutrient cations differ from those of anions.

The significance of micronutrients has become increasingly apparent over the last three decades, especially as micronutrient deficiencies have been observed in many intensively cultivated soils across the country. This deficiency is largely attributed to several factors in modern Indian agriculture: (a) continuous depletion of micronutrients from soils due to high-yield crop varieties, particularly cereals, which require and respond to substantial fertilizer input, leading to reduced micronutrient levels in soil solutions; (b) the widespread use of high-analysis fertilizers lacking micronutrients; and (c) advances in instrumentation and a deeper understanding of plant mineral nutrition that now allow for more precise identification of micronutrient deficiencies that were previously unnoticed. To meet the rising demand for crop production, it is essential to understand soil factors that influence the supply and availability of micronutrients.

Micronutrient source in soil

Micronutrients in soils primarily exist as oxides, sulphides, and silicates, originating from soil-forming rocks and minerals that transform soil development. The micronutrient levels in soil are influenced by the chemical makeup of its parent material. Both deficiencies and toxicities of micronutrients in the soil can result from either low or excessive concentrations in the original rocks and minerals from which the soil has formed. Micronutrient distribution within soils is highly variable, and spatial differences can be substantial. Natural sources of soil micronutrients differ significantly among soil types, and a general overview of these variations for Indian soils is provided in Table 1. In addition to inorganic minerals, organic matter plays an important role in contributing to the total micronutrient content in surface soils. Furthermore, soil amendments and fertilizers are common additions to the micronutrient reserves in soils.

Table 1: Major natural sources of micronutrients and the content in surface soils

Micronutrient	Major forms in nature	Total content in surface soil (ppm)	Available content in surface soil (ppm)*
Zinc	Sulphides, oxides, carbonates, silicates	7 to 1000	0.8-20.5
Copper	Sulphides, oxides, hydroxy carbonates, silicates	1.8 to 960	Trace 32.0
Iron	Sulphides, oxides, silicates	4000 to 273000	0.36-174
Manganese	Oxides, carbonates, silicates	37 to 115000	0.60-164
Boron	Borates, borosiliates	2.8 to 630	Trace 24**
Molybdenum	Sulphides, oxides, molybdates	Trace to 12	Trace 2.8***
Chlorine	Chlorides	20 to 1000	0.5-10**
Nickel	Silicates	-	-
Cobalt	-	4 to 78	0.2-0.5

*DTPA-$CaCl_2$ extractable; **Hot water extractable; ***Ammonium oxalate extractable

During pedogenesis, extensive transformations occur in soil minerals, resulting in the release of micronutrients. Micronutrient cations such as Fe^{2+}/Fe^{3+}, Zn^{2+}, Mn^{2+}, and Cu^{2+} are liberated through these processes and are subsequently either adsorbed onto soil colloids or incorporated into secondary silicate minerals via isomorphous substitution. Similarly, micronutrient anions such as molybdates (MoO_4^{2-}) and borates (BO_3^{3-}) react chemically with soil constituents, including phosphates, to form stable reaction products. Chloride (Cl^-) is present in significant concentrations in soils, with additional inputs from precipitation. Micronutrients are predominantly retained in tightly bound organic and inorganic forms, becoming available to plants only gradually through chemical weathering and microbial decomposition mechanisms.

Micronutrients status in Indian soils

The total concentrations of Fe, Mn, Zn, Cu, Co, Cl, and B exhibit considerable variation across diverse soil types in India (Table 1). With the exception of Zn, Cu, and B, the majority of micronutrients are present in adequate quantities to sustain agricultural productivity. However, the total contents of Zn, Cu, B, and Mo are comparatively lower in light-textured soils of northern India, particularly in Punjab, Haryana, Delhi, Uttar Pradesh, and Uttarakhand, as well as in the sandy and alluvial soils of western India, notably Gujarat. Boron concentrations are also significantly deficient in eastern India, especially in the acidic soils of Bihar, Jharkhand, West Bengal, Assam, and Odisha, where widespread deficiencies have been reported. The available micronutrient content of Indian soils, determined using DTPA + TEA + $CaCl_2$ (pH 7.3) for Zn, Cu, Fe, Mn, and Co, reveals substantial variability (Table 1). Except for Zn, the DTPA-extractable concentrations of other micronutrients generally remain within satisfactory ranges to support optimal crop growth. Zinc deficiency, however, is prevalent across all agroecological zones, irrespective of soil type or cropping system. The deficiency is particularly severe in soils with a pH $\geq$ 8 and in coarse-textured soils. Furthermore, Zn deficiency is observed in nearly all rice-growing regions of the country. Boron is the second most deficient micronutrient in Indian soils after Zn, with significant deficiencies reported in the acidic soils of eastern India, including Bihar, Jharkhand, West Bengal, Assam, and Odisha. Boron deficiency is also observed in many soils of Uttar Pradesh and Uttarakhand.

Physiological functions of micro-nutrients

Micronutrient cations play a critical role as cofactors in various enzyme systems. Except for Zn, these cations also function as electron carriers within enzymes involved in oxidation-reduction reactions in plants. Their physiological roles can be summarized as follows:

- Zinc, copper, and manganese are integral to carbohydrate and protein metabolism through their involvement in multiple enzyme systems. Zinc, in particular, plays a key role in the biosynthesis of certain growth-promoting hormones and is essential for reproductive processes in many plant species, which are crucial for grain development. Key metalloenzymes associated with Zn include carbonic anhydrase, dehydrogenases, proteinases, and peptidases.
- Enzymes containing specific micronutrients play critical roles in plant physiological processes:
- Iron is a component of enzymes such as cytochrome oxidase, catalase, and peroxidase.
- Copper is a key element in enzymes like cytochrome oxidase, ascorbic acid oxidase, and polyphenol oxidase.
- Manganese is integral to dicarboxylases and dehydrogenases involved in the tricarboxylic acid (TCA) cycle. It is also an essential element in Photosystem II, where it facilitates photolysis.
- Molybdenum is crucial for the activity of nitrogenase and nitrate reductase enzymes, making it essential for nitrogen fixation in plants and microorganisms.
- Cobalt is indispensable for symbiotic nitrogen fixation, supporting the growth of rhizobia and the formation of the coenzyme cobalamin.
- Nickel is associated with various enzyme systems and plays a regulatory role in plant metabolism.
- Boron influences water absorption and the translocation of sugars within plants.

Thus, micronutrients are vital for numerous physiological processes, contributing significantly to the growth and development of plants and microorganisms.

Micronutrient pools in soil and their significance in plant uptake

The total micronutrient content in soil is distributed across various organic and inorganic fractions, each with distinct characteristics. These fractions do not correspond to classical chemical compounds, and their availability to plants varies significantly. Currently, there is no universally accepted method for quantifying the magnitude of these fractions. Micronutrient cations in soil are primarily found in five fractions: (i) water-soluble, (ii) exchangeable, (iii) complexed, adsorbed, or chelated, (iv) occluded and bound to carbonates, sesquioxides, or secondary minerals, and (v) primary minerals. Numerous studies have attempted to fractionate the various pools of soil zinc (Zn) to quantify these fractions and assess their relative importance for plant uptake. Research indicates that over 95% of soil Zn is associated with primary minerals, which contribute minimally to plant availability. Conversely, less than 5% of soil Zn is present in water-soluble, exchangeable, and complexed fractions, yet these forms are the most significant for plant uptake. When zinc fertilizers are applied, 12–34% of the added Zn typically remains in the water-soluble, exchangeable, and complexed fractions, depending on soil properties. This fraction is crucial as it contributes to the residual availability of zinc fertilizers. The distribution patterns of other micronutrients across different fractions follow a similar trend, with only a small portion in readily available forms contributing significantly to plant nutrition.

Soil conditions favouring micronutrient deficiency and toxicity

Certain soil characteristics can lead to either deficiencies or toxicities of micronutrients, adversely affecting crop growth. The following conditions are known to cause micronutrient deficiencies in soils:

1. Soils with a very high pH (alkaline soils).
2. Light-textured soils with low pH (acidic soils) that develop under regions with high annual rainfall.
3. Soils subjected to intensive cultivation and application of high-analysis NPK fertilizers for high-yielding crop varieties.
4. Organic soils (peat or muck soils)

Soils derived from parent materials inherently low in micronutrients are more prone to micronutrient deficiencies, particularly under conditions of acid leaching or intensive cropping. Acidic soil conditions are especially favourable for molybdenum (Mo) deficiency. Conversely, high pH soils, such as sodic soils, commonly exhibit deficiencies of zinc (Zn), iron (Fe), manganese (Mn), and copper (Cu). Additionally, over liming of acidic soils can lead to a reduction in the availability of micronutrient cations. In acid sulphate soils, micronutrient toxicity often restricts crop growth under upland conditions. However, in waterlogged soils, such as those used for rice cultivation, these issues are mitigated due to an increase in soil pH. Iron toxicity in rice is frequently observed in soils with high Fe content under highly reducing, submerged conditions, especially in the presence of easily decomposable organic matter. Such conditions can also lead to Fe-induced Zn deficiency. Soils near copper and zinc mining areas, often exposed to mine effluents, may exhibit micronutrient toxicity that adversely affects crop growth. In organic soils, micronutrient deficiencies, particularly of Cu, can limit plant growth as micronutrient cations tend to bind strongly with organic ligands, reducing their availability to plants.

Factors influencing the availability of micronutrient cations and anions

Micronutrient cations, such as copper (Cu), iron (Fe), manganese (Mn), zinc (Zn), cobalt (Co), and nickel (Ni), are positively charged, while boron (B), molybdenum (Mo), and chlorine (Cl) exist as anions with negative charges. The influence of soil conditions on the availability of micronutrient cations differs significantly from that of anions. Therefore, the factors affecting the availability of these two groups of micronutrients will be examined separately.

1. *Effect of Cations*

Several soil factors influence the availability of all micronutrient cations in a similar manner. These factors are discussed in the following sections.

a) Soil pH: Soil pH is a critical factor that governs the solubility and availability of micronutrient cations in the soil. Extremely low pH indicates highly acidic conditions, where micronutrient cations are generally found in the soil solution. In such environments, some of these cations may become toxic to plants, and the presence of a dominant ion in the soil solution can impede the uptake of other micronutrient cations, especially if they are present in low concentrations. Conversely, when soil pH exceeds 7.0, micronutrient cations tend to precipitate as insoluble hydroxides. Under alkaline conditions, these cations are primarily converted into oxides or hydroxides with lower solubility. Examples of such behaviour are detailed for each micronutrient cation.

$$Fe^{2+} + 3OH^- \longrightarrow Fe(OH)_2$$

$$Cu^{2+} + 2OH^- \longrightarrow Cu(OH)_2$$

$$Mn^{2+} + 2OH^- \longrightarrow Mn(OH)_2$$

$$Zn^{2+} + 2OH \longrightarrow Zn(OH)_2$$

$$\text{(Soluble ion)} \longrightarrow \text{(Insoluble compound)}$$

The hydroxides of micronutrient cations are generally poorly soluble, but their solubility and availability vary. Each micronutrient cation has a distinct pH range for the formation of its hydroxide. Additionally, the oxidation state of the micronutrient cation influences the formation of precipitates in relation to soil pH. Insoluble hydroxides begin to form when the pH of the soil solution reaches 6.0 or higher. As the pH increases beyond 6.0, the solubility and bioavailability of micronutrient cations decrease, highlighting the importance of maintaining a soil pH around 6.0 to keep these cations in adequate concentrations in the soil solution. Liming is often applied with the aim of maintaining a pH of approximately 6.0 to prevent micronutrient cation concentrations from reaching toxic levels. However, excessive liming can lower the concentration of these cations in the soil to a point where sensitive plants may experience micronutrient deficiencies.

b) Oxidation States of Micronutrient Cations: Micronutrient cations in lower oxidation states, such as manganese (Mn^{2+}) and copper (Cu^{2+}), are generally more soluble than those in higher oxidation states under typical soil pH conditions. The reduced forms are favored in environments with limited oxygen, such as waterlogged soils. Under high moisture conditions, the presence of easily decomposable organic matter promotes anaerobic conditions, supporting microbial activity that derives energy from the oxidation of inorganic ions. A common example is the conversion of divalent Mn^{2+} to trivalent Mn^{3+} by certain fungi and bacteria. Soil pH plays a significant role in the redox state of micronutrient cations. High pH promotes oxidation, while low pH conditions favor the reduction of these cations. For instance, at very low pH (around pH 3.0), ferric iron (Fe^{3+}) precipitates as hydroxide, while ferrous iron (Fe^{2+}) remains in solution until the pH rises above 6.0. At higher oxidation states, the hydroxides of micronutrient cations become extremely insoluble and largely unavailable to plants. In well-drained, aerated calcareous soils, micronutrient cations predominantly exist in their oxidized states, leading to low availability despite potentially high total micronutrient content. In such soils, the hydroxides and carbonates of these cations are too insoluble to meet plant requirements. For example, zinc, though a divalent cation, forms zincate ions $Zn(OH)_4^{2-}$ at high pH. These micronutrient cations can also exist in various complexed forms, such as $ZnCl_2$, $Zn(OH)_4^{2-}$, $Fe(OH)_3$, and $Cu(OH)_2$.

c) Interaction of Micronutrient Cations with Soil Constituents: The availability of micronutrient cations in soil is significantly influenced by the presence of various soil constituents, including inorganic ions in the soil solution, free oxides of iron and aluminium, soil organic matter, and applied fertilizers or amendments. Silicate clays, especially 2:1 type clay minerals, can contain micronutrient cations such as zinc (Zn), iron (Fe), and manganese (Mn) within their octahedral layers. These ions can be released from the clay under certain conditions or adsorbed onto the colloidal surfaces when their concentration in the soil solution increases due to micronutrient fertilizer application. Such adsorption can lead to the fixation of micronutrients, particularly zinc and copper, posing a challenge for their availability, while the high total content of iron and manganese in soils makes their fixation less problematic. The presence of carbonate and bicarbonate ions in soils, which may arise from sodicity or excessive liming, reduces the availability of micronutrient cations for crops, especially for orchard crops and field crops that are prone to iron deficiency. Additionally, low soil temperatures can hinder the availability of micronutrient cations. For instance, wheat growers in northern India often observe zinc deficiency in their crops when soil temperatures drop due to freezing. This deficiency is primarily caused by the reduced solubility of micronutrient cations under low temperature conditions. Furthermore, wet soil conditions in cold temperatures exacerbate the issue, further limiting the availability of micronutrient cations. The capacity of soil organic matter to form stable complexes with micronutrient cations is well documented. A variety of organic compounds, including humic and fulvic acids, as well as other organic acids, play a role in the formation of organo-metallic complexes with varying solubilities. Some metal ions are bound as insoluble complexes, such as those with humic acids, making them unavailable to plants. However, certain organic ligands can form soluble chelates with micronutrient cations, which are then available for plant uptake. Microorganisms also assimilate these micronutrient cations, as they are essential for many microbial metabolic processes. These metal ions are temporarily immobilized in microbial biomass but are released back into the soil during the mineralization of dead microorganisms. During the decomposition of plant and animal residues, microorganisms produce chelating agents that convert insoluble micronutrient cations into soluble metal complexes, thus enhancing their availability to plants. Metal chelates are relatively stable and protected from soil reactions, allowing for their efficient uptake by plants. Natural chelating agents are found in the leachates of forest soil humus layers, and metal complexes with fulvic acids are particularly water-soluble, making them accessible to plants. Organic soils can retain significant amounts of micronutrient cations in insoluble forms through complexation. Copper is strongly retained in organic soils, and these insoluble complexes contribute to its low availability. Similarly, the availability of zinc is also reduced in organic soils due to comparable mechanisms.

2) Effect of cations

Chlorine, boron, and molybdenum exist as anions in soils, and their behavior in the soil environment differs significantly from one another.

a) Chlorine: Among the three anions, chloride is absorbed by plants in significant quantities. Its availability in soil remains largely unaffected by soil properties, as most chloride compounds are highly water-soluble. A considerable amount of chloride enters the soil through rainfall. In humid regions, chloride is prone to leaching due to its lack of adsorption by soil colloids, resulting in its movement deeper into the soil profile. Conversely, in arid and semi-arid regions, surface soils often exhibit high chloride concentrations, particularly in saline soils, where it may even reach levels toxic to certain crops. Chlorine deficiency in soils is uncommon, as chloride is also introduced into soils through impurities present in various agrochemicals.

b) Boron: The availability of boron in soil is significantly influenced by soil pH. At lower pH levels, most boron compounds are soluble, and boron is readily available to plants in the form of boric acid. In sandy soils with low pH, high rainfall can lead to substantial boron leaching, resulting in its loss from the soil profile. However, in fine-textured soils, boron leaching is less problematic unless the pH is extremely low. At low pH, boron is not strongly adsorbed or fixed by soil constituents. As soil pH increases, boron availability declines. Liming reduces boron availability, primarily due to lime-induced fixation of boron by clay and other minerals, as soluble forms like sodium and calcium borates become less prevalent. Overliming has been associated with boron deficiency in crops, as evidenced by widespread deficiencies in the calcareous soils of northern Bihar. Boron is predominantly found in the organic matter of surface soils, with its concentration decreasing down the soil profile. The mineralization of organic matter releases boron, making it available for plant uptake. During drought conditions, boron deficiency in plants is often observed due to reduced boron availability in subsoil layers.

c) Molybdenum: Molybdenum availability in soils closely resembles that of phosphorus, as both are influenced by similar soil factors. Conditions that enhance phosphorus availability also improve the availability of molybdenum. In acidic soils, molybdenum availability is significantly reduced because it becomes fixed in iron and aluminium compounds and on silicate surfaces. Phosphate application can release fixed molybdate ions through anion exchange, suggesting that molybdate and phosphate share the same fixation sites on soil colloids. Liming acidic soils effectively increases the availability of native molybdenum. In regions like Australia and New Zealand, liming is commonly practiced, as it can achieve similar yield improvements in legumes as the application of molybdenum fertilizers. In areas where liming is cost-prohibitive, applying molybdenum fertilizers alongside phosphatic fertilizers has shown economic benefits, especially in legume and vegetable crops. Field experiments have demonstrated that the combined application of molybdate and phosphate positively impacts their uptake in molybdenum-deficient soils. Conversely, the presence of sulphate reduces the availability and uptake of molybdenum by crops, though the exact mechanisms behind this interaction are not yet fully understood.

Soil management to maintain adequate levels of micronutrients

Effective soil management to maintain adequate levels of micronutrients requires a thorough understanding of the interactions between macro-, secondary, and micronutrients, as well as the relationships among various micronutrients themselves. Both synergistic and antagonistic interactions between micronutrients have been well-documented. For instance, the beneficial effects of molybdenum (Mo) and phosphorus (P) in legumes, manganese (Mn) and Mo in field crops, and zinc (Zn) and P at optimal application levels in field and orchard crops are well established. Additionally, adequate calcium (Ca) levels in soil are crucial for improving boron (B) uptake. Proper balance is also necessary between potassium (K) and copper (Cu), as well as K and iron (Fe), to ensure normal nutrient absorption. Similarly, sufficient Zn levels are vital for the optimal utilization of Cu and Mn. Recent studies have focused on the antagonistic interactions between nutrient ions, which can significantly affect plant nutrient uptake. Understanding the mechanisms underlying these micronutrient interactions in soils and plant systems is essential for developing effective management strategies.

Interactions of micronutrients

The availability of sufficient levels of micronutrients in the soil alone does not guarantee adequate micronutrient nutrition for plants. In addition to soil factors, plant growth factors significantly influence the uptake and assimilation of micronutrients. These processes are also strongly impacted by interactions between different micronutrients, as well as with macronutrients and secondary nutrients. Such interactions can occur within the soil, at the root-soil interface, within the plant itself, or across all these locations. These interactions can alter plant nutrition and must be thoroughly understood to implement effective agronomic practices aimed at achieving optimal crop yields.

Interaction mechanism: Nutrient interaction refers to the influence of one element on another concerning plant growth and crop yield. This is typically evaluated by applying two nutrients simultaneously at varying levels and observing the differential response to one element in the presence of the other. For instance, the yield response from individual applications of zinc or iron in rice may differ from the response when both zinc and iron are applied together. When the combined application of two nutrients results in a yield response greater than the sum of their individual responses, the interaction is considered positive. Conversely, if the combined effect leads to a yield response lower than the sum of the individual responses, the interaction is deemed negative. Positive interactions among micronutrients enhance crop yields and should be utilized to boost agricultural production. On the other hand, negative interactions can reduce yields and should be carefully avoided when developing agronomic strategies.

It is crucial to understand the mechanisms behind both positive and negative interactions, whether they occur within the plant or in the surrounding environment. For example, the negative interaction between phosphorus (P) and zinc (Zn) has been extensively studied. This interaction may negatively impact plant growth and yield due to several possible mechanisms, as discussed below:

(i) Chemical Precipitation of Zinc in Soil: The formation of insoluble zinc phosphates, such as $Zn_3(PO_4)_2$ and $Zn_3(PO_4)_2.4H_2O$, was initially believed to cause phosphorus-induced zinc deficiency in soils. However, solubility studies on these compounds have demonstrated that they release sufficient zinc into the soil solution to support healthy crop growth. Therefore, it is now evident that phosphorus-induced zinc deficiency in plants is not due to the chemical precipitation of zinc by phosphorus in the soil.

(ii) Reduction of Zinc Translocation in Plants: Research has shown that phosphorus inhibits the translocation of zinc within plant roots, leading to lower zinc concentrations in the aerial parts of the plant and reduced zinc uptake. Phosphorus application appears to interfere with the uptake, movement, and assimilation of zinc by plants. It has been observed that phosphorus and zinc exhibit mutual antagonism when either nutrient exceeds a critical threshold within the root system. However, the precise chemical or physiological mechanisms underlying phosphorus-induced interference with zinc in plant roots remain unclear.

(iii) Dilution Effect: The dilution effect occurs when the rate of dry matter accumulation exceeds the rate of nutrient uptake, resulting in a decrease in the nutrient's concentration within the dry matter. Zinc deficiency symptoms in plants may emerge due to this effect during periods of rapid growth. This phenomenon is commonly observed when phosphorus fertilizer is applied to soils deficient in phosphorus but only marginally deficient in zinc, where zinc fertilizer is typically not added. Phosphorus application enhances plant growth rates, but the uptake of zinc may not increase proportionally to sustain adequate zinc concentrations in the above-ground plant parts. This imbalance can lead to the manifestation of zinc deficiency symptoms in plant tissues. The severity of the dilution effect is influenced by the extent of the plant's response to phosphorus fertilization and the rate at which zinc concentrations decline in the plant's dry matter. In cases where the response to phosphorus is minimal or negligible, and the available zinc in the soil is marginal, the dilution effect on zinc uptake may not be evident, even with high phosphorus application rates.

(iv) Uneven Distribution of zinc in Roots and Shoots: Zinc exhibits lower mobility within plants compared to phosphorus, resulting in significant accumulation of zinc in tissues such as leaves, nodes, and internodes. This effect is particularly evident in soils with low levels of available zinc. When small amounts of phosphorus fertilizer are applied, zinc uptake and concentration tend to increase in the roots while simultaneously decreasing significantly in the shoots. Research suggests that the antagonistic interaction between phosphorus and zinc occurs within the root system. However, the precise mechanism underlying this negative interaction remains unclear. Current evidence indicates that it is neither purely chemical nor biochemical, nor strictly physiological, as no conclusive proof has yet been established. The phosphorus-induced zinc deficiency in plants, often associated with reduced yield, can be mitigated by applying zinc fertilizer alongside phosphorus fertilizer. This practice transforms the negative interaction into a positive one, as zinc fertilizer helps maintain adequate zinc levels in leaves and other plant tissues, thereby improving plant growth and yield.

In addition to the interactions between phosphorus (P) and zinc (Zn), various interactions occur between micronutrients and macronutrients, as well as among different micronutrients. These interactions can often be explained by the mechanisms previously described. The following antagonistic effects on micronutrient uptake by crops have been well-documented:

- High levels of phosphorus reduce the utilization of zinc, iron (Fe), and copper (Cu).
- Excessive iron negatively impacts the uptake of zinc and manganese (Mn).
- Elevated concentrations of zinc, manganese, and copper can lead to iron deficiency in crops.
- Sulphur (S) and copper in excess can induce molybdenum (Mo) deficiency.
- High lime content in soil causes deficiencies in all micronutrients.

Management of micronutrients in soil-plant system

The examples of both beneficial and antagonistic relationships among nutrients highlight the complexities involved in managing soils to maintain adequate levels of micronutrients. As agricultural production intensifies to maximize yield per unit area, it becomes crucial to ensure balanced nutrient availability to address nutritional limitations that may restrict crop productivity. The specific micronutrient requirements of crops, combined with the intricate interrelationships among these nutrients, make it challenging to establish generalized soil management practices for maintaining optimal micronutrient levels. Furthermore, since the margin between toxicity and deficiency levels for many micronutrients is narrow, tailored, location specific expertise and management strategies are essential to ensure effective and sustainable micronutrient utilization.

Understanding the factors influencing the availability of micronutrients in soils provides valuable insights for developing effective management practices to maintain them in plant-available forms. Key considerations include soil texture, pH, water regime, and calcium carbonate content, which should be carefully evaluated before implementing any management strategies. Additionally, the intensity of cropping systems and the pattern of fertilizer use play a significant role. In systems with high NPK fertilizer consumption, micronutrient deficiencies are more likely to occur. Micronutrient fertilizers should be recommended based on soil test results indicating deficiencies. Emerging micronutrient deficiencies can be addressed by selecting crop varieties capable of efficiently absorbing micronutrients from soil solid phases. A straightforward approach to managing such deficiencies involves the periodic application of micronutrient fertilizers alongside NPK fertilizers, preferably in alternate years, as indicated by soil test results. Soil application of micronutrient fertilizers tends to be more effective for field crops, whereas for fruit trees, both soil and foliar applications are considered effective. Foliar application can also be cost-efficient when micronutrients are combined with plant protection chemicals and macronutrients.

Micronutrients are often present in acid soils in excessive amounts, which can become toxic to sensitive crops. Once micronutrients are introduced into the soil system, they are challenging to remove. However, their solubility and availability can be effectively managed through the application of suitable soil amendments. For example, adding lime to acidic soils significantly reduces the availability of micronutrients. Conversely, in calcareous, alkaline, and sodic soils, deficiencies of micronutrients such as iron (Fe), zinc (Zn), copper (Cu), and manganese (Mn) are commonly observed. The use of amendments like gypsum, pressmud, or sulphur has been shown to lower soil pH and enhance the availability of these micronutrients. Incorporating green manure is also an effective strategy for improving the fertility of such soils, especially when cultivating rice under submerged conditions. Research indicates that maintaining soil pH around 6.0 is optimal for ensuring an adequate supply of micronutrients in the soil, supporting balanced nutrient availability for crop growth.

The availability of micronutrients in soil is significantly influenced by the soil moisture regime. Under well-drained conditions, soils maintain an aerobic environment, causing micronutrients to predominantly exist in their oxidized states. In these oxidized forms, micronutrients are generally less available, even in acidic soils, compared to their reduced forms. Conversely, in submerged soil conditions, the reduced forms of iron (Fe) and manganese (Mn) dominate the soil solution. However, zinc (Zn) and copper (Cu) may become deficient in such conditions if their native concentrations in the soil are insufficient to meet crop requirements. A modern approach to micronutrient management involves cultivating crop varieties capable of extracting essential micronutrients from insoluble sources in the soil. This can be achieved through genetic engineering, which combines traits for efficient nutrient utilization with high-yield potential. Such genetically enhanced crop varieties are already being cultivated in regions like Australia and the United States, eliminating the need for micronutrient fertilizer application in some cases.

Soil tests for micronutrients

The primary purpose of micronutrient soil testing is to determine whether a field is deficient or adequate in a specific micronutrient. This information is essential for implementing corrective measures to address micronutrient deficiencies effectively. It also helps the fertilizer industry develop region-specific micronutrient fertilizers tailored to local soil conditions. Furthermore, government agencies rely on soil test data to plan and execute agricultural development programs. Another critical goal of these tests is to identify fields where the application of micronutrient fertilizers would result in economically viable yield improvements.

Soil testing involves four key steps:

1) Collection of representative and uncontaminated soil samples: Ensuring that samples accurately reflect the field's soil conditions.
2) Selection of appropriate methods for extracting and determining available micronutrient content: Employing reliable and standardized procedures to assess soil nutrient availability.
3) Calibration of soil test results with crop responses: Establishing the relationship between soil nutrient levels and plant growth or yield.
4) Interpretation of soil test results and providing recommendations: Offering scientifically sound guidance for nutrient management based on test findings.

Each step in the soil testing process is crucial and demands a high level of expertise in micronutrient chemistry and the behavior of micronutrients within soil and plant systems. Consequently, soil testing for micronutrients has evolved into a specialized field requiring in-depth knowledge. Micronutrient soil tests are primarily designed to assess the magnitude of micronutrient pools in the soil that are linked to plant availability. Efforts have also been made to develop soil testing methods capable of estimating bioavailable micronutrients using a single extract. The extraction agents employed in these tests fall into distinct categories, including water, dilute acids, chelating or complexing agents and neutral buffered or unbuffered salts. These chemicals are utilized either individually or in combination to enhance their efficiency in extracting micronutrients from soil samples. The most commonly employed soil tests for assessing micronutrient availability are outlined in Table 2.

Table 2: Soil test methods for micronutrients

Element (ppm)	Extractant	Soil:Solution	Shaking period (min)	Deficiency critical level
Zinc	0.005M DTPA+0.1M TEA*+0.01M $CaCl_2$ (pH 7.3)	1:2	120	0.5-1.0
	0.1 N HCL	1:5	30	1-5
	1 N NH_4OAc (pH 4.6)	1:10	60	0.2-0.5
	0.05 N HCl	1:2	5	1.0
Copper	0.005M DTPA+0.1M TEA*+0.01M $CaCl_2$ (pH 7.3)	1:2	120	0.2-0.5
	0.1 N HCL	1:5-10	30	1-3
	1 N NH_4OAc (pH 4.6)	1:2-4	60	0.2
Iron	0.005M DTPA+0.1M TEA*+0.01M $CaCl_2$	1:2	120	2.5-5.8
	1 N NH_4OAc (pH 4.8)	1:10-20	60	2
Manganese	0.005M DTPA+0.1M TEA*+0.01M $CaCl_2$	1:2	120	2-4
	1 N NH_4OAc (pH 7.0)	1:10	30	3-4
	3 N $NH_4H_2PO_4$ and 0.1 N H_3PO_4	1:10	60	15-20
Boron	Boiling hot water	5:10	5	0.5-1.0
Molybdenum	Ammonium oxalate (pH 3.3)	1:10	360	0.05-0.2

*TEA: Triethanol amine

Each soil testing method varies in terms of shaking time, soil-to-solution ratio, and critical limits. These critical limits for different extractants can differ even for the same micronutrient, depending on the soil conditions and crop type. Climate also plays a significant role in determining these critical limits for micronutrients. It is important to note that soil test values are empirical and may not directly correlate with the deficiency or sufficiency of a micronutrient under specific soil conditions unless they are properly calibrated with crop response. Therefore, it is crucial to consider various soil factors when interpreting soil test results. These factors must be carefully considered during interpretation. After thoroughly assessing the impact of these factors, soil tests can provide accurate predictions on the amount of micronutrient fertilizer needed to achieve the maximum economic benefit for crop production.

Common micronutrient carriers and their rates of application

Both soluble and insoluble micronutrient sources are commonly used in agricultural practices to ensure adequate micronutrient levels in the soil. A list of commonly used micronutrient fertilizers, along with their recommended application rates, is provided in Table 5. For each micronutrient, there are one or two primary fertilizer carriers typically employed. In India, the most widely used fertilizer for zinc is zinc sulphate heptahydrate ($ZnSO_4.7H_2O$), whereas in the USA and other countries, zinc sulphate monohydrate and zinc oxysulphate are preferred due to their higher concentration compared to zinc sulphate heptahydrate. Zinc oxide (ZnO) is another zinc fertilizer source, though its zinc content can vary due to impurities. For other micronutrient cations, their corresponding sulphates are typically used as fertilizers because they are water-soluble. Manganese dioxide (MnO_2) is also used as a fertilizer, though it is less efficient than manganese sulphate ($MnSO_4$). Micronutrient chelates, which are water-soluble, are becoming increasingly popular for use in field and orchard crops. Boron is typically applied as sodium tetraborate or boric acid, both of which are water-soluble and readily available. Molybdenum is commonly applied as ammonium molybdate or sodium molybdate, which are also water-soluble. Chlorine fertilizers, such as sodium chloride and calcium chloride, are used depending on the crop, particularly in coconut cultivation.

Micronutrient management in soil requires specialized knowledge. While various commercial preparations are available for foliar application, they are often ineffective in soils with micronutrient deficiencies. Proper soil management and soil application of the correct micronutrient levels are more beneficial for crop growth. Unlike macronutrients such as nitrogen (N), phosphorus (P), potassium (K), and sulphur (S), excessive application of micronutrients can negatively impact plant growth, degrade soil fertility, and reduce crop yields. Effective soil management requires a deep understanding of micronutrient behavior under different soil conditions.

10

Soil Fertility Evaluation

Soil fertility is defined by the soil's inherent capacity to supply essential nutrients in adequate quantities and proper balance to support the growth of specific plants, provided that factors such as light, moisture, temperature, and soil physical conditions remain favorable. It is characterized as a component of the soil-plant relationship, emphasizing plant growth in connection with nutrient availability in the soil. In 1840, the "Law of Restitution" was introduced by Justus Von Liebig, asserting that soil fertility must be maintained by replenishing the nutrients removed by crops through the application of manures and fertilizers. Soil fertility evaluation is conducted to assess the nutrient-supplying capacity of the soil, which involves understanding several critical concepts that influence soil fertility. Additionally, the "Law of Minimum" was proposed by Von Liebig, stating that crop growth will be restricted by the availability of the single nutrient that is insufficient, even if all other essential nutrients are present in adequate amounts.

Soil fertility evaluation approaches

Soil fertility evaluation methods can be categorized into the following approaches:

1. Soil Analysis
2. Plant Analysis
3. Biological Methods
4. Visual Symptoms of Nutrient Deficiency or Toxicity

Soil testing and plant analysis are valuable tools for making fertilizer application recommendations. Soil testing assesses the availability of nutrients to crops, while plant analysis provides insight into the actual nutrient uptake from the soil.

The soil testing objectives include:

1. Categorizing soils based on nutrient levels to guide fertilizer recommendations.
2. Estimating the likelihood of obtaining profitable crop responses.
3. Assisting in evaluating soil productivity.
4. Identifying specific soil conditions such as alkalinity, salinity, and acidity that may limit crop yields.

Available nutrients: Plants obtain their nutrients from air, water, and soil, with the majority of mineral nutrients coming from the soil. The soil's available nutrient forms are those present in various discrete chemical states, often existing in a dynamic equilibrium, from which plants can absorb them. The available fraction of a nutrient is the one whose variation significantly impacts crop yield and responses. The nutrients that are accessible to biological organisms are referred to as bioavailable nutrients.

Chemical methods for nutrient estimation

The chemical methods for nutrient estimation include soil testing, which involves measuring the availability of nitrogen (N), phosphorus (P), potassium (K), sulphur (S), micronutrients, lime and gypsum requirements, along with assessing pH, electrical conductivity (EC), calcium carbonate content, and soil texture using the Bouyoucos hydrometer method.

Table 1: Different extractants for available nutrients

Nutrient	Extractant
Available P	0.5 M $NaHCO_3$, pH 8.5 Olsen's extractant
	0.03 N NH_4F + 0.025 NHCl Bray's No. 1 extractant
Available K	Neutral normal ammonium nitrate
Available S	0.15% $CaCl_2$
Available Zn, Fe, Cu, Mn	0.005 M DTPA, pH 7.3 Diethylene Triamine penta acetate
Gypsum requirement	Schoonover method

Table 2: Rating limits of soil test values

Nutrient	Low	Medium	High
Organic carbon (%)	Below 0.5	0.5-0.75	Above 0.75
Available N (kg/ha)	Below 280	280-560	Above 560
Available P (kg/ha)	Below 10	10-24.6	Above 24.6
Available K (kg/ha)	Below 108	108-280	Above 280

Available S (ppm)	Deficient < 10 ppm	Sufficient > 10 ppm
DTPA Zn	Deficient < 0.6 ppm	Sufficient > 0.6 ppm

Visual Indicators of Nutrient Deficiency or Toxicity

Plants require a balanced and consistent supply of 16 essential nutrients, as established by the criteria of essentiality, to ensure proper growth and development. Although chemical analysis of plants may detect the presence of more than 90 elements, these 16 nutrients are crucial for plant health. Imbalances in nutrient levels, whether due to deficiencies or toxicities, can result in nutritional disorders. When an essential nutrient is in extremely limited or excessive supply, plants exhibit specific signs, known as deficiency or toxicity symptoms. Deficiency symptoms are often distinct and can be identified based on the affected plant part and the growth stage at which they appear. The mobility of nutrients within a plant determines the location of these symptoms. Mobile nutrients such as nitrogen (N), phosphorus (P), and potassium (K) are readily translocated from older to younger leaves under stress conditions, causing deficiency symptoms to initially appear on older leaves. In contrast, immobile nutrients such as calcium (Ca), sulphur (S), boron (B), and iron (Fe) cannot be retrains located, and their deficiency symptoms first manifest on younger leaves. The mobility of some other nutrients falls between these two extremes. For example, zinc (Zn) deficiency can impair auxin metabolism, leading to shortened internodes. The visual identification of nutrient deficiencies or toxicities is a practical and cost-effective diagnostic approach in plant nutrition, as it does not require specialized analytical equipment.

Limitations

- Accurate diagnosis using this approach requires substantial expertise, as certain nutrient deficiency symptoms can be challenging to distinguish from one another.
- By the time deficiency or toxicity symptoms become visible, the crop may have already suffered significant setbacks. Interventions applied at this stage may not fully restore potential yields. The manifestation of visible symptoms represents an advanced stage of nutrient deficiency. However, even in the absence of overt symptoms, crop yield may decline due to insufficient nutrient levels—a condition often referred to as "hidden hunger."

Plant analysis

Plant analysis, though an indirect method of assessing soil fertility, serves as an important complement to soil testing. It is particularly useful for verifying nutrient deficiencies, toxicities, or imbalances; detecting hidden hunger; evaluating fertilizer programs; and determining nutrient availability. In some cases, nutrients may be present in sufficient quantities in the soil, but factors such as inadequate soil moisture or imbalances with other nutrients may limit their availability to plants. This method operates on the principle that the concentration of a specific nutrient in plant tissues reflects the supply of that nutrient in the soil and is directly correlated to its availability. For diagnostic purposes, plant analysis is often interpreted using the critical value approach. This method establishes a threshold nutrient concentration in plant tissues, typically calibrated to represent 90–95% of the maximum potential yield. Values below this threshold indicate deficiency, while those above signify sufficiency. Plant analysis generally falls into two categories:

1. Tissue Testing: Conducted on fresh plant tissues directly in the field.
2. Total Analysis: Performed in laboratories using precise analytical techniques.

1. Tissue Tests

Tissue testing involves rapid assessments to determine nutrient elements in the sap extracted from fresh plant tissues. These tests evaluate the sap from ruptured cells for unassimilated nitrogen, phosphorus, and potassium. They are semi-quantitative methods primarily used to confirm or predict deficiencies of these nutrients. The results are typically categorized as low, medium, or high. Proper application of tissue tests can help anticipate or address production issues while crops are still in the field. The nutrient concentration in cell sap provides a reliable indication of the plant's nutrient status at the time of testing.

1. Plant Part Selection: Generally, the conductive tissue of the most recently matured leaf is selected for testing.
2. Optimal Testing Time: The critical stage for tissue testing is from bloom to the early fruiting stage. Nutrient concentrations, such as nitrates, tend to be higher in the morning compared to the afternoon when supplies are limited.

Common Tests:

- Nitrates: Diphenylamine test
- Phosphates: Molybdate and stannous oxalate test
- Potassium: Sodium cobaltinitrite test

2. Total Analysis

Total analysis involves assessing the nutrient content of entire plants or specific plant parts. This process employs precise analytical methods to measure various elements after the plant material is dried, ground, and ashed. It is used to estimate the total nutrient content present in the plant. The percentage of average plant nutrient concentrations (by weight on dry matter basis) is as follows:

- Hydrogen: 6.0%
- Oxygen: 45.0%
- Carbon: 45.0%
- Nitrogen: 1.5%
- Potassium: 1.0%
- Calcium: 0.5%
- Magnesium: 0.2%
- Phosphorus: 0.1%
- Sulphur: 0.1%
- Chlorine: 100 ppm (0.01%)
- Iron: 100 ppm
- Boron: 20 ppm
- Manganese: 50 ppm
- Zinc: 20 ppm

- Copper: 6 ppm
- Molybdenum: 0.1 ppm

Critical Nutrient Concentration (CNC)

CNC is defined as the nutrient level below which yield of crop, quality, or performance becomes unsatisfactory. However, determining an exact threshold can be challenging, as it varies with crops and conditions. For instance, in crops like sugar beet, an excessive nitrogen concentration can adversely impact quality. In such cases, the CNC is interpreted as a maximum rather than a minimum value. Therefore, it is often more practical to consider the Critical Nutrient Range (CNR)—a specified range of nutrient concentrations at a particular growth stage. The CNR defines the range above which the crop is sufficiently supplied with nutrients and below which deficiencies may occur.

Diagnosis and Recommendation Integrated System (DRIS)

Diagnosis and Recommendation Integrated System introduced by Beaufils in 1973, evaluates plant tissue composition by analyzing nutrient concentration ratios rather than individual nutrient levels. This method assesses the relative balance of nutrients using index values, where negative values indicate deficiencies, and positive values reflect sufficiency. DRIS not only identifies the most limiting nutrient but also ranks nutrients in the order of their likelihood to become limiting. By addressing all nutritional constraints, this system enhances fertilizer recommendations and improves the likelihood of achieving high crop yields. Index values are calibrated to measure deviations of specific nutrients in plant tissues from optimal levels, enabling the classification of yield-limiting factors in order of significance. Developing a DRIS for a particular crop requires meeting the following criteria:

1) Clearly identifying all factors potentially affecting crop yield.
2) Establishing the relationship between these factors and crop yield.
3) Defining calibrated norms for the crop.
4) Continuously refining recommendations to suit specific conditions, ensuring accurate and effective use of the established norms.

Development of DRIS Norms: A large number of randomly selected sites are chosen to represent the entire production area. At each site, plant and soil samples are collected for analysis of all essential elements. The entire dataset is divided into two subgroups: high and low yielders, based on factors such as plant vigor, quality, and yield. Nutrient concentrations in the plants are expressed in various forms, such as nutrient ratios (e.g., N/P, N/K) or products of nutrients (e.g., N×P, N×K). Any form of expression that effectively differentiates between the high and low yielding subgroups is retained as a useful diagnostic parameter. The mean values for each of these expressions across all sites are then used to establish the diagnostic norms. The crop's NPK requirements are assessed using a DRIS chart. This chart is constructed with three axes representing N/P, N/K, and K/P ratios, each using mean values from the high-yielding subgroup. Concentric circles on the chart represent confidence limits, with horizontal arrows in the inner circle indicating nutrient balance. Diagonal arrows indicate nutrient imbalances, and the inner circle is set at +15%, while the outer circle is at the mean plus 30% for each expression. Vertical arrows indicate a specific nutrient imbalance. These arrow notations can be replaced with DRIS indices to further refine the analysis.

1. The significance of maintaining nutritional balance is emphasized.
2. The established norms for elemental content can be applied universally.
3. Diagnoses can be made across a wide range of growth stages.
4. Nutrients that limit yield due to either excess or deficiency can be easily identified.

Indicator Plants

Some plants are highly sensitive to the deficiency of a particular nutrient, displaying unique symptoms that distinguish them from other deficiency symptoms. This allows for the straightforward detection of the nutrient's deficiency. The plants that serve as indicator plants is presented below:

- Nitrogen: Cauliflower, Cabbage
- Phosphorus: Rapeseed
- Potassium: Potato
- Calcium: Cauliflower, Cabbage
- Magnesium: Potato
- Iron: Cauliflower, Cabbage, Potato
- Sodium: Sugar beet
- Manganese: Sugar beet, Oats, Potato
- Boron: Sunflower

Critical levels of different nutrients in soil and plants

Concentration or ranges of the major elements and micronutrients in mature leaf tissue generalized as deficient, sufficient or excessive for various plant species

Essential elements	Deficient	Sufficient or normal	Excessive or toxic
Major elements (%)			
Nitrogen	< 2.50	2.50-4.50	> 6.00
Phosphorus	< 0.15	0.20-0.75	> 1.00
Potassium	< 1.00	1.50-5.50	> 6.00
Calcium	< 0.50	1.00-4.00	> 5.00
Magnesium	< 0.20	0.25-1.00	> 1.50
Sulphur	< 0.20	0.25-1.00	> 3.00

Essential elements	Deficient	Sufficient or normal	Excessive or toxic
		Micronutrients (ppm)	
Boron	5-30	10-200	50-200
Chlorine	< 100	100-500	500-1000
Copper	2-5	5-30	20-100
Iron	< 50	100-500	> 500
Manganese	15-25	20-300	300-500
Molybdenum	0.03-0.15	0.1-2.0	> 100
Zinc	10-20	27-100	100-400
		Soil characteristics	
	Low	**Medium**	**High**
Organic carbon (%)	< 0.4	0.4-0.75	> 0.75
Phosphorus (ppm)	< 5.0	5-9	9-20 > 20 (very high)
Potash	< 55.0	-	> 55

11

Fertilizer Use Efficiency

Fertilizer Use Efficiency (FUE) is a critical agronomic concept that quantifies the effectiveness of fertilizer applications in enhancing crop productivity while minimizing nutrient loss and environmental degradation. Fertilizers are essential for replenishing soil nutrients, particularly nitrogen (N), phosphorus (P), and potassium (K), which are vital for plant growth. However, the efficiency with which these nutrients are absorbed by crops and utilized can vary significantly depending on several factors, including soil properties, crop species, climatic conditions, and fertilizer management practices. FUE is typically assessed through several indicators, including nutrient uptake efficiency, nutrient recovery efficiency, and agronomic efficiency. Nutrient uptake efficiency refers to the proportion of available nutrients absorbed by the plant relative to the total nutrients applied, while nutrient recovery efficiency measures the fraction of applied nutrients that are effectively utilized by the crop. Agronomic efficiency, defined as the increase in crop yield per unit of fertilizer applied, is another key metric for evaluating fertilizer effectiveness.

Inefficient fertilizer use not only leads to suboptimal crop growth but also contributes to environmental concerns, such as nutrient leaching, greenhouse gas emissions, and eutrophication of water bodies. Therefore, improving FUE is crucial for sustainable agricultural practices, ensuring that fertilizers are applied in a manner that maximizes plant uptake, minimizes losses to the environment, and reduces input costs for farmers. Strategies to enhance FUE include precision nutrient management approaches such as Site-Specific Nutrient Management (SSNM), the use of Enhanced Efficiency Fertilizers (EEFs), and advanced soil testing techniques. Integrated Nutrient Management (INM), which combines the use of organic and inorganic fertilizers, crop rotation, and organic amendments, also plays a significant role in improving nutrient use efficiency. Optimizing fertilizer application, while considering the physiological demands of the crop and the specific characteristics of the soil, is vital for achieving both high agricultural productivity and environmental sustainability. The factors that enhance fertilizer use efficiency are listed below:

1. Irrigated crops generally show a stronger response to fertilizers compared to rainfed crops. Therefore, farmers should prioritize fertilization for crops grown in fields with controlled irrigation over those in unirrigated fields.

2. To maximize the benefits of limited fertilizer resources, farmers should select the most responsive and suitable crop varieties for their specific locality.
3. High-yielding crop varieties tend to respond better to fertilizers, even at lower application rates, compared to local varieties. Hence, when feasible, high-yielding varieties should be preferred.
4. Delayed sowing significantly reduces both the yield and the efficiency of fertilizer use. For example, in regions like Punjab, Haryana, and Western UP, when rice planting is delayed beyond July 15, the yield decreases by 2 to 3 quintals per hectare for each additional day of delay. To optimize fertilizer benefits, crops should be planted at the appropriate time for the respective region.
5. Ensuring appropriate plant spacing is crucial for optimal crop yields. Typically, higher yields are observed when fertilizer is applied at similar rates with closer row spacing. Ideal spacing includes 15 cm between rows and 10 cm between plants for kharif rice, 10 cm by 10 cm for rabi rice, and 22.5 cm between rows for wheat.
6. It is important to effectively manage and utilize organic waste materials on the farm to improve soil fertility, enhance water retention capacity, and provide plant nutrients as a supplement to chemical fertilizers. These organic materials should be incorporated into the soil 3 to 4 weeks before sowing. Applying farmyard manure during the kharif season is preferable. Given that cow dung is often used as fuel, small-scale biogas plants can meet household fuel needs while ensuring the manure is fully utilized as an organic fertilizer, retaining its nutrient value.
7. When using compost made from straw and leaves, which typically have a high carbon-to-nitrogen (C:N) ratio, it is necessary to add small amounts of nitrogenous fertilizers to the compost to prevent nitrogen deficiency for young crops. This is especially important when the crop is planted shortly after compost application. Incorporating finely ground rock phosphate or superphosphate with organic manures can help create a more balanced nutrient profile for the crops.
8. The incorporation of leguminous crops into diverse crop rotations and intercropping systems, along with the use of bacterial and algal cultures, plays a vital role in meeting the nutrient requirements of crops. Leguminous plants are unique in their ability to fix atmospheric nitrogen in the soil, with rates ranging from 50 to 250 kg N per hectare per season. Additionally, they can mobilize unavailable native soil phosphorus and convert it into an accessible form. Research has shown that the inclusion of pulse crops in rotations enhances nitrogen uptake by subsequent cereal crops, while cereal-dominant rotations, such as rice-rice-rice or maize/jowar/bajra-

wheat-maize/bajra (fodder), show lower nitrogen uptake. It is essential to inoculate leguminous crop seeds with the appropriate *Rhizobium* culture. For inoculation, mix 100 g of gum arabic, 225 ml of water, and 150 g of peat or lignite-based *Rhizobium* inoculant. This mixture is sufficient to treat 7 kg of small seeds or 14 kg of medium-sized seeds, followed by uniform coating with neutralizing agents like calcium carbonate, ground rock phosphate, or gypsum for acidic soils.

9. Whenever feasible, green manuring with dhaincha (*Sesbania aculeata*) or sunhemp (Crotalaria juncea) should be practiced in lowland paddy cultivation to improve soil fertility.
10. Irrigation practices should be carefully managed, providing just enough water at different physiological growth stages, as recommended for the specific crop. Excessive irrigation can lead to significant nutrient losses, especially nitrogen (N) and potassium (K) through leaching. In waterlogged conditions, a substantial amount of nitrogen is lost in gaseous form. Therefore, it is important to avoid excess water in the soil during and immediately after fertilizer application. This is particularly true for urea, which should only be applied on a drained field (moist soil). Subsequent irrigation should be carried out 24 to 48 hours after fertilization. Proper drainage of excess water is just as crucial as irrigation during moisture-deficit conditions.
11. Balanced fertilization, ensuring adequate availability of nutrients (including micronutrients) from both the soil and applied fertilizers, should be practiced based on soil testing.
12. Fertilizer recommendations should ideally consider the crop sequence, particularly in multiple cropping systems. Farmyard manure is best applied during the wet season, while phosphate fertilizers are more effective during the dry season. Intercropping systems involving crops like green gram, black gram, soybean, onion, and groundnut are highly profitable in wide-row crops such as cotton and red gram. These intercrops do not require additional fertilizers and provide a better economic return compared to monoculture.
13. Phosphorus and potassium fertilizers should primarily be applied as basal dressings, with adjustments based on soil texture. In light-textured soils, a portion of the potash may be top-dressed. Nitrogen, however, should be applied in two to three split doses, depending on the crop's duration, growth stages, and soil moisture availability. For rice, 50% of the nitrogen should be applied as a basal dose, 25% at the tillering stage, and the remaining 25% at the panicle initiation stage. In medium to high fertility soils or when nitrogen fertilizer is limited, a 50-50 split between the tillering and panicle initiation stages is recommended. If a single nitrogen application is used, it

should be top-dressed 30–35 days after planting and incorporated into the soil.

14. For irrigated wheat, nitrogen application varies based on soil type. In the heavy black soils of Madhya Pradesh, a single basal dose is effective, whereas, on the light soils of Punjab and Rajasthan, splitting the nitrogen in a 1:1 ratio at sowing and the first irrigation gives the best results. In regions like Himachal Pradesh and the Tarai areas, where the growing season is extended, nitrogen should be applied in a 1:1:1 ratio at sowing, first irrigation, and late tillering stages.
15. For maize hybrids, nitrogen is most effective when applied in three splits: one-third at planting, and one-third each at 30 and 40 days after planting. Similarly, for kharif jowar, half the nitrogen should be applied at sowing, and the rest at the knee-high stage. In potato cultivation, nitrogen should be split equally between planting and within 30 days of planting, before stolon formation.
16. Phosphorus fertilizers should be placed 4–6 cm below and 4–6 cm away from the seeds to ensure their maximum availability to the crops. Potassium fertilizers can be placed alongside phosphorus or broadcast and mixed thoroughly into the soil. In dry soils, fertilizers must be placed within the moist zone. Nitrogen and potassium fertilizers applied as a top dressing should be incorporated into the top 10 cm of the soil for optimal nutrient use.
17. The best results from urea application are achieved when it is pre-treated with moist soil before top-dressing. Mixing one part of urea with 5–10 parts of moist soil and curing the mixture for about 24 hours significantly reduces nitrogen losses and enhances its effectiveness.
18. Zinc deficiency has become increasingly prevalent in agricultural soils. To address this, zinc sulphate should be applied to the soil as a basal application at a rate of 25–50 kg per hectare. For standing crops exhibiting zinc deficiency symptoms, a foliar spray of 0.3–0.5% zinc sulphate solution mixed with 0.3% lime solution can provide quick relief and recovery.
19. Rock phosphates are cost-effective source of phosphorus and can be used efficiently in acidic soils to enhance crop productivity.
20. In acidic, saline, or alkali soils, the application of appropriate soil amendments such as lime or gypsum is essential before using chemical fertilizers. For alkali soils, gypsum should be broadcast at a rate of 3–5 tonnes per acre (8–12 tonnes per hectare) and thoroughly mixed into the top 10 cm of the soil layer to improve soil structure and nutrient availability.
21. Weeds, especially during the initial growth stages of crops in the kharif season (7–21 days), compete aggressively for nutrients, potentially reducing the availability of applied nutrients by 25–30%. Therefore, effective weed

management during the early stages of crop growth is critical to ensure optimal nutrient use efficiency.

22. Effective control of pests and diseases is essential to maximize the efficiency and benefits of fertilizer application. Unchecked pest or disease infestations can compromise crop growth and reduce the overall nutrient uptake, limiting the effectiveness of the applied fertilizers.

12

Method and Time of Fertilizers Application

To achieve maximum benefits from fertilizers, it is crucial to apply them at the correct time and location. Fertilizers differ in their solubility in water and mobility within the soil. Similarly, soils vary in texture, ranging from sandy to clayey, which affects the movement and availability of applied nutrients. Furthermore, the nutrient requirements of plants differ across their growth stages. For instance, nitrogen is absorbed throughout the entire growth cycle, whereas phosphorus uptake is most rapid during the early stages of growth. Consequently, the timing and method of fertilizer application must consider:

1. The characteristics of the fertilizer.
2. The type of soil.
3. The nutrient requirements and growth patterns of specific crops.

In addition to these factors, several general principles guide the selection of the optimal time for fertilizer application.

Principles for Determining the Appropriate Time for Fertilizer Application

1. Nitrogen is essential throughout the growth cycle of a crop. Plants typically grow at a slower pace during the initial stages, followed by a phase of rapid growth known as the grand growth period, and then a gradual slowdown as they approach maturity. Correspondingly, nitrogen uptake mirrors this growth pattern: it is minimal during the early stages, peaks during the grand growth phase, and decreases as the plant matures. Thus, nitrogen demand is sustained throughout the growth period, but its requirement varies with the growth stage.
2. Nitrogenous fertilizers are highly water-soluble, making them mobile and capable of rapid movement in all directions within the soil. This characteristic also makes nitrogen prone to significant losses through leaching.

   Since nitrogen is required continuously throughout the crop growth period, and its loss through leaching is inevitable, it is advisable to avoid applying large quantities of nitrogenous fertilizers at once. Instead, splitting the application into multiple doses during the growth cycle ensures a steady supply of nitrogen to the plants. This approach prevents nitrogen deficiency at critical growth stages and optimizes nutrient availability. These principles are effectively implemented in nitrogen fertilizer recommendations for sugarcane, a crop with high nitrogen demand. For sugarcane planted in October-November (pre-seasonal) and June-July (adsali) in Maharashtra, the recommended nitrogen rates are 420 kg/ha and 504 kg/ha, respectively. The application of nitrogenous fertilizers is carried out in split doses according to a carefully designed schedule tailored to the specific planting seasons, such as those followed in the Deccan Canal region of Maharashtra (Table 1).

Table 1: Application time of fertilizers for different plantings of sugarcane in Maharashtra

	October-November (pre-seasonal planting) 420 kg N/ha		**June-July (adsali planting) 5054 kg N/ha**	
Time of sowing	Ammonium sulphate Kg/ha	Groundnut cake	Ammonium sulphate	Groundnut cake
At planting	34	332	41.5	-
6-8 weeks after	23	-	26.4	393
12-16 weeks after	34	-	26.4	-
At earthing-up	23	332	26.4	393

Directorate of Publicity, Government of Bombay, Bombay, 1957

The significance of split application of nitrogenous fertilizers in wheat cultivation has been well-documented. Tandon (1982) reported that across 23 experiments conducted at 10 locations, splitting the nitrogen application into two doses—half at sowing and the other half during the first irrigation (3-4 weeks after sowing)—proved highly effective. Similarly, Rathi and Vaishya et al. (1983) demonstrated that applying 90 kg N/ha in three splits (30 kg as a basal dose, 30 kg at 30 days after sowing, and 30 kg after the first cutting) maximized green fodder yield in oats.

3. Phosphorus, on the other hand, plays a critical role during the early stages of root development and initial plant growth. Approximately two-thirds of the total phosphorus requirement is absorbed by crop plants when they have accumulated just one-third of their total dry weight. This highlights the importance of providing an adequate phosphorus supply during the early growth stages to support optimal development.
4. All phosphatic fertilizers release phosphorus for plant uptake gradually. Even superphosphate, which contains water-soluble monocalcium phosphate (P_2O_5), undergoes partial conversion upon application to the soil. The water-soluble phosphorus is transformed into dicalcium phosphate or citrate-soluble forms, which are less soluble and slowly available to plants. Since phosphorus is required in significant amounts during the early stages of plant growth, but its release from fertilizers is slow, it is generally recommended to apply the entire dose of phosphatic fertilizers as a basal application before sowing or planting.

5. Potassium exhibits characteristics similar to both nitrogen and phosphorus. In terms of uptake, it behaves like nitrogen, being absorbed continuously throughout the growing season until harvest. However, potassic fertilizers, like phosphatic fertilizers, release nutrients gradually. Therefore, it is advisable to apply the entire quantity of potassic fertilizers at the time of sowing to ensure a steady supply of potassium throughout the growth period.
6. Leaching losses are more pronounced in sandy soils compared to heavier clayey soils. This necessitates more frequent or split applications of nitrogenous fertilizers, and occasionally potassic fertilizers, in sandy soils to minimize nutrient losses and maintain their availability to crops.

Practical recommendations for the timing of fertilizer application, derived from these principles, are summarized as follows:

1. Nitrogenous fertilizers should be applied in split doses, with two splits for crops of 4–5 months duration, three splits for crops of 9–12 months duration, and four to five splits for long-duration crops, such as the adsali sugarcane crop.
2. On sandy or light-textured soils, more frequent or split applications of nitrogenous fertilizers are recommended compared to heavy clay soils, to minimize leaching losses.
3. The full dose of water-soluble phosphatic fertilizers should be applied as a basal application at the time of sowing. For acid soils, bone meal or rock phosphate should be applied 1–2 weeks prior to sowing for optimal results.
4. Potassic fertilizers should also be applied entirely at the time of planting.

Principles for Selecting Appropriate Fertilizer Application Methods

1. Nitrogenous fertilizers are highly water-soluble and exhibit rapid movement in all directions from the point of application. This characteristic allows them to quickly reach plant roots when applied on the soil surface. Therefore, they are commonly broadcast on the soil just before sowing.
2. Due to the susceptibility of nitrogen to leaching, it is applied at different growth stages of the crop. The rapid mobility of nitrogenous fertilizers in moist soil ensures that surface application followed by irrigation adequately meets the plant's nitrogen demand during critical growth phases. Consequently, nitrogenous fertilizers are well-suited for top-dressing and side-dressing applications.
3. Since phosphorus moves slowly within the soil, it must be placed in close proximity to the plant roots to ensure accessibility.
4. The progressive fixation of phosphates by soil clays reduces their availability over time. Fixation refers to chemical or physical interactions between applied nutrients and the soil that render them less accessible to plants. To

minimize phosphate fixation, phosphatic fertilizers should be placed to limit their contact with soil particles while being closer to plant roots.

This localized placement near seeds or seedling roots is preferred for three key reasons:

(i) Limiting the contact of fertilizers with the soil reduces the fixation of phosphates.

(ii) Essential nutrients are positioned within easy reach of plant roots, and the risk of harmful concentrations is minimized when placement is precisely controlled.

(iii) Fertilizers applied in a side band along crop rows do not readily supply nutrients to weeds growing between the rows.

Since potassic fertilizers exhibit slow mobility in soil, they should be positioned close to the root zone for effective uptake.

Methods of Fertilizer Application

A.Application of Fertilizers in Solid Form

1. Broadcasting
   (i) Broadcasting at planting
   (ii) Top-dressing
2. Placement
   (i) Plough-sole placement
   (ii) Deep placement
   (iii) Sub-soil placement
3. Localized placement
   (i) Contact placement or Combined drilling or Drill placement
   (ii) Band placement
   (iii) Pellet application
   (iv) Side-dressing

B.Application of fertilizers in liquid form

1. Starter solutions
2. Foliar application or Spray fertilization
3. Direct application to the soil
4. Application through irrigation water

A. Application of Fertilizers in Solid Form

1. Broadcasting

In the broadcasting method, fertilizers are uniformly distributed across the entire soil surface to ensure even application. This can be carried out before ploughing, just before planting, or during the crop's growth stage. Depending on the timing, broadcasting is classified into two types: broadcasting at planting and top-dressing. For densely planted crops such as wheat and barley or crops sown through broadcasting, spreading fertilizers on a standing crop is referred to as top-dressing.

(i) Broadcasting at Planting

The primary objectives of broadcasting fertilizers at planting are:

a) To achieve uniform distribution of fertilizers and incorporate them into the plough layer or soil surface.
b) To allow for the application of larger quantities of fertilizer than can safely be applied during planting using seed and fertilizer drills.

Conditions Favoring Broadcasting at Planting:

a) When nitrogenous fertilizers are required for soils with severe nitrogen deficiency or after nutrient-depleting crops such as fodder sorghum and maize. Nitrogenous fertilizers like ammonium sulphate, ammonium sulphate-nitrate, and concentrated organic manures are suitable for this purpose.
b) When citrate-soluble phosphatic fertilizers (e.g., basic slag, dicalcium phosphate) or insoluble phosphatic fertilizers (e.g., bone meal, ground rock phosphate) need to be applied to moderately to strongly acidic soils.
c) When potassic fertilizers such as murate of potash or potassium sulphate are required for soils deficient in potassium.

(ii) Top-dressing

Top-dressing involves applying nitrogenous fertilizers, such as sodium nitrate, calcium ammonium nitrate, ammonium sulphate-nitrate, and ammonium nitrate, to closely spaced crops like paddy and wheat. This method provides nitrogen in a readily available form to support the plants' growth. Several top-dressings may be made during critical growth stages when crops have the highest nitrogen demand. Phosphorus and potassium top-dressing is typically limited to pasture lands that remain under cultivation for extended periods. When performing top-dressing, it is crucial to avoid applying fertilizers when plant leaves are wet, as this can cause burning or scorching. This damage is more significant with nitrogenous and potassic fertilizers than with phosphatic fertilizers. In certain regions, airplanes are used for fertilizer application in hilly terrains where transport is difficult or where large quantities are needed due to severe nutrient deficiencies. Aerial application offers several advantages, including:

1. Application of small quantities of fertilizers over large areas, such as for micronutrients.
2. Use of high-analysis fertilizers.

3. Integration with other aerial operations, such as insect control.
4. Time and labor savings.

For effective aerial top-dressing, an airstrip near the fields and suitable weather conditions are essential. In areas like Assam and Nagaland, top-dressing with basic slag or ground rock phosphate can help improve soil fertility in extensive, acidic lands. Numerous trials in Assam's Darrang district have demonstrated that phosphatic fertilizers significantly enhance paddy yields. Given the acidic nature of soils in Assam and Nagaland, applying insoluble phosphatic fertilizers can be particularly beneficial.

In New Zealand, the application of superphosphate using aircraft is a widespread practice. By 1956, approximately 38.4 million hectares were top-dressed via aerial methods. Similarly, in the United States, significant areas of paddy fields are sown and top-dressed annually with the help of aircraft. In New Zealand, aerial application of superphosphate is a widely practiced method. By 1956, approximately 38.4 million hectares of land were top-dressed using aircraft. Similarly, in the United States, significant areas of paddy fields are sown and fertilized annually through aerial methods.

2. Placement

In placement methods, fertilizers are incorporated into the soil regardless of the positioning of seeds, seedlings, or growing plants. This can be done either before or after sowing the crops. The following are some commonly adopted placement techniques:

(i) Plough-Sole Placement: This method involves placing fertilizers in continuous bands at the bottom of the furrow during ploughing. Each fertilizer band is subsequently covered as the next furrow is turned. The positioning of the crop is not specifically aligned with the fertilizer bands. This technique is particularly useful in areas where the soil tends to dry out a few inches below the surface during the growing season or in soils with a dense clay pan just below the plough layer. Fertilizers are positioned in the moist soil, ensuring they remain available to the plants even during dry periods. Additionally, localized placement using this method reduces the fixation of phosphorus and potassium compared to broadcasting, as it minimizes the interaction between fertilizer and soil particles.

(ii) Deep Placement of Nitrogenous Fertilizers: Deep placement of nitrogenous fertilizers is widely practiced in paddy fields, especially in Japan, and is also recommended for use in India. This method involves applying ammoniacal fertilizers like ammonium sulphate or ammonia-forming fertilizers like urea in the reduction zone of the soil. In this zone, the fertilizers remain in their ammonia form, ensuring a sustained supply of nitrogen to the crop during its active vegetative growth phase. Deep or subsurface placement enhances nutrient distribution within the root zone while preventing losses through surface runoff. The specific technique for deep placement depends on local cultivation practices. In irrigated areas with reliable water supply, fertilizers are applied under the plough furrow in dry soil before flooding the fields. For regions where puddling is practiced, fertilizers are broadcast before puddling, which subsequently incorporates them deeper into the root zone, ensuring efficient nutrient utilization by the plants. A series of experiments conducted at the Central Rice Research Institute, Cuttack (Odisha), demonstrated the advantages of deep placement of ammonium sulphate over conventional surface application. Surface application of ammonium sulphate at a rate of 22.4 kg of nitrogen per hectare increased paddy yield by 87.5 kg. In contrast, deep placement of the same dose, 7.5 cm below the surface, resulted in an additional yield of 133.3 kg. These findings highlight the enhanced efficiency of ammonium sulphate when applied through deep placement.

Similarly, the application of superphosphate during puddling improved yields from 0.9 to 3.4 g per pot. However, when superphosphate was deep-placed, the yield increased further to 5.9 g per pot, indicating a significant improvement in phosphate efficiency with deep placement.

(iii) Sub-Soil Placement: Sub-soil placement involves applying fertilizers directly to the subsoil using heavy machinery. This method is particularly suitable for humid and sub-humid regions where subsoils are strongly acidic and deficient in available plant nutrients. Under these conditions, the placement of fertilizers, particularly phosphatic and potassic types, in the subsoil facilitates improved root development and nutrient uptake. In recent years, there has been growing interest in the United States regarding the potential benefits of sub-soil fertilization in regions with nutrient-deficient subsoils.

3. Localized Placement

This method involves applying fertilizers directly into the soil near seeds or plants, ensuring a localized nutrient supply. Localized placement is typically used when small quantities of fertilizers need to be applied. By placing fertilizers close to the seed or plant in bands or pockets, young plants have access to an adequate nutrient supply, promoting early and vigorous growth. This accelerated growth facilitates early intercultural operations, which improve weed control. Furthermore, localized placement minimizes phosphorus and potassium fixation, enhancing nutrient availability. Several techniques are employed to place fertilizers near seeds or plants, including;

(i) Contact Placement or Combined Drilling

Contact placement, also known as drill placement, involves sowing seeds and fertilizers simultaneously in the same row. This method is particularly effective for applying phosphatic and potassic fertilizers to crops such as cereals (wheat, barley, sorghum, pearl millet, and finger millet), cotton, and grasses. However, excessive fertilizer concentrations in close proximity to seeds can hinder germination or damage young plants, especially in dry, sandy soils. Nitrogen and potassium fertilizers tend to increase soil solution salinity more than phosphate fertilizers. Consequently, it is crucial to apply only small amounts of fertilizers to prevent adverse effects on germination.

(ii) Band Placement

In this method, fertilizers are applied in bands, which may be continuous or discontinuous, near the seed or transplanted plant. Band placement can be categorized into two types:

a) Hill Placement: Hill placement involves placing fertilizers in bands near individual plants when they are spaced at least 90 cm apart. The length, depth, and distance of the band from the plant depend on the crop and the fertilizer quantity applied. This method is commonly used in India for applying nitrogenous and phosphatic fertilizers to fruit trees such as oranges, bananas, papayas, apples, pears, coconuts, and cashew nuts. In states like Gujarat and Maharashtra, hill placement is practiced for applying nitrogenous fertilizers and oil cakes to American cotton grown under irrigated conditions, spaced at 7.5 × 7.5 or 10 × 10 cm. Hill placement is also widely adopted by vegetable growers.
b) Row Placement: Row placement involves placing fertilizers in continuous bands on one or both sides of a row of closely spaced plants. This can be done manually or with a seed drill. Crops such as sugarcane, potato, maize, tobacco, cotton, cereals, and vegetables commonly benefit from this method. Row placement allows for higher rates of fertilizer application compared to hill placement. However, for smaller quantities of fertilizer, hill placement tends to be more efficient.

(iii) Pellet Application

Pellet application is a localized fertilization technique where nitrogenous fertilizers are formed into pellets and placed 2.5 to 5 cm deep in the soil. This method involves mixing the fertilizer with soil in a 1:10 ratio, forming the mixture into a dough, shaping it into small pellets, and depositing them into the soft mud of paddy fields between the crop rows. Research conducted at the Central Rice Research Institute, Cuttack (Odisha), demonstrated that applying nitrogen in pellet form significantly improves its efficiency, more than doubling the nitrogen utilization compared to the conventional top-dressing method commonly practiced by farmers.

(iv) Side-Dressing

Side-dressing involves applying fertilizers around the plants or between crop rows during the growth period. This term encompasses various practices aimed at providing supplemental nutrients to crops. The commonly used side-dressing methods include:

a) Row Application for Nitrogenous Fertilizers: Nitrogenous fertilizers are manually applied between rows of broad-row crops such as maize, sugarcane, tobacco, sorghum, barley, cauliflower, cabbage, tomato, brinjal, and cotton. This practice supplies additional nitrogen to support the crop's growth during critical stages.

(v) Application Around Fruit Trees: Mixed or straight fertilizers are applied around the base of fruit trees like papaya, banana, orange, grape, apple, mango, and guava. The frequency of application varies based on the tree's age, ranging from once to three times annually. This practice, sometimes referred to as hill application, ensures a steady nutrient supply for optimal growth and fruit production.

The choice of fertilizer application methods depends on the crop type, nutrient requirements, and local conditions. For field crops, phosphorus and potassium are typically combined with a portion of nitrogen and applied through methods like combine-drilling or side-dressing at sowing. Additional nitrogen, on the other hand, is better applied later in the growth period through broadcast or side-dressing methods. Employing a combination of methods and timings for fertilizer application has been shown to enhance nutrient use efficiency compared to relying on a single method or application time.

B. Application of liquid fertilizers

The application of liquid fertilizers has gained prominence as an effective fertilization technique. There are four main methods of applying liquid fertilizers:

1. Starter Solutions: Starter solutions are dilute mixtures of fertilizers, typically containing nitrogen (N), phosphorus (P_2O_5), and potassium (K_2O) in ratios such as 1:2:1 or 1:1:2. These solutions are applied to young vegetable plants during transplanting, replacing the customary watering to aid plant establishment. The amount of fertilizer used in starter solutions is minimal. This method offers two primary advantages: (i) nutrients are immediately available to plant roots, and (ii) the diluted solution prevents any negative effects on growth. Consequently, starter solutions promote rapid establishment and early growth. However, the method has two limitations: (i) it requires additional labor if watering is not part of the routine operation, and (ii) it may lead to increased phosphate fixation.

2. Foliar Application: Foliar application involves spraying fertilizer solutions directly onto the leaves of growing plants. These solutions are prepared in low concentrations and can supply either a specific nutrient or a combination of nutrients. Foliar application is especially beneficial for addressing specific nutrient deficiencies and supplementing soil-applied fertilizers. It is well-established that plant nutrients can be absorbed through foliage, with some nutrients being absorbed both rapidly and efficiently. Studies indicate that the effects of nutrient sprays are comparable to soil applications, with efficacy rarely exceeding two to three times that of soil-applied fertilizers. Consequently, foliar application does not significantly reduce fertilizer usage but may be preferred under specific conditions:
   - When soil conditions or competition from other crops limit nutrient availability, such as during late nitrogen application in rain-fed farming systems.
   - When a precisely timed fertilizer response is needed, for example, in response to seasonal changes.
   - When routine spraying of pesticides or herbicides is combined with nutrient application, a practice becoming popular among orchardists in the United States.
   - When crop growth restricts soil application but allows foliar application, using high-clearance sprayers or helicopters.

   Despite its advantages, foliar application has certain limitations:
   - Marginal leaf burn or scorching may occur if strong solutions are used.
   - Since low-concentration solutions (typically 3–6%) are required, only small quantities of nutrients can be applied in a single spray.
   - Achieving moderate to high fertilizer rates necessitates multiple applications.
   - Foliar application is generally more expensive than soil application unless combined with other spraying operations, such as for pest, weed, or disease control.
3. Direct Application of Liquid Fertilizers to Soil: Specialized equipment is used to directly inject liquid fertilizers, such as anhydrous ammonia (a liquid under high pressure, up to 14 kg/cm^2 or more), into the soil. This technique is particularly popular in the United States. When applied approximately 10 cm below the seed, this method minimizes plant injury and nitrogen loss. However, shallow applications can lead to nitrogen losses as ammonia escapes. This approach allows for the efficient use of cost-effective nitrogen sources. In Mysore, an indigenous bullock-drawn implement has been successfully utilized for injecting anhydrous ammonia into the soil. This device consists of a gas cylinder mounted on wheels,

which supplies ammonia under controlled pressure into a furrow created by a type of coulter. Similarly, in Japan, simple equipment has been developed for the application of ammonia solutions. For optimal results with anhydrous ammonia application, the soil should have moisture at field capacity, and the soil tilth should be ideal. Medium-textured soils are particularly suitable as they provide the widest range of moisture levels for effective application. However, significant nitrogen losses can occur in sandy or clayey soils that are either excessively dry or overly wet. In addition to liquid nitrogen-based fertilizers, multi-nutrient liquid fertilizers containing nitrogen (N), phosphorus (P), and potassium (K) are directly applied to soils in the United States.

4. Application of Liquid Fertilizers through Irrigation Water: Water-soluble fertilizers, either straight or mixed formulations containing N, P, and K, are dissolved in the irrigation stream, allowing nutrients to be distributed into the soil in solution. This method reduces application costs and enables the use of cost-effective, water-soluble fertilizers. Nitrogenous fertilizers are most commonly applied through this method due to their solubility and effectiveness.

References

Tandon, P. (1982). Optimal patents with compulsory licensing. *Journal of Political Economy*, *90*(3), 470-486.

Vaishya, U. K., G. R. Gayendragadkar, and R. L. Pandey. (1983) Effect of Rhizobium inoculation on nodulation and grain yield of mungbean (*Vigna radiata* (L) wilczek): 228-230.

13

Organic Manure

Historically, the term "manure" referred to organic materials such as animal dung and other bulky natural substances used to enhance crop production. Over time, the scope of materials used for crop enhancement expanded to include chemical substances, such as ammonium sulphate $(NH_4)_2SO_4$ and superphosphate, which supply essential nutrients to plants. Consequently, the distinction emerged: organic substances came to be specifically referred to as manures, while inorganic substances were termed fertilizers.

Definition

Manures are naturally derived plant and animal residues utilized as a source of nutrients for plant growth. Alternatively, manures can be defined as organic materials, either bulky or concentrated, that supply essential plant nutrients while also improving the physical properties of the soil. These materials do not have a fixed chemical composition.

Advantages of Organic Manures

1. Supply of nutrients: Provide primary, secondary, and micronutrients essential for plant growth.
2. Soil improvement: Enhance the physical structure and condition of the soil.
3. Pest and disease management: Aid in controlling plant parasitic nematodes and fungal pathogens.
4. Carbon dioxide enrichment: Act as a source of CO_2 for plants, contributing to photosynthesis.
5. Support for soil microbes: Serve as a food source for beneficial soil microorganisms, promoting biological activity.
6. Buffering capacity: Help regulate soil pH by providing buffering action.

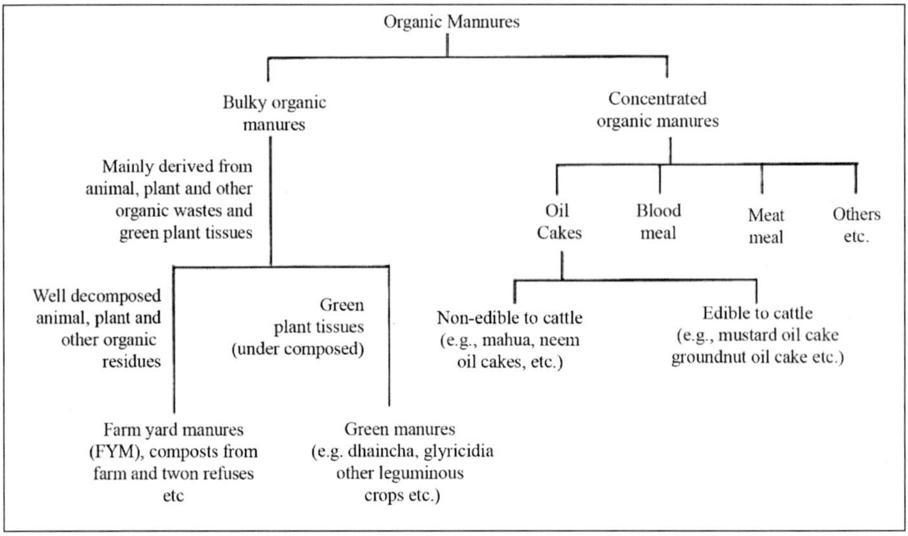

A. Bulky organic manure

The category of bulky organic manures includes materials such as farmyard manure (F.Y.M.), farm manure, town compost, compost, green manures, night-soil, sludge, and other similar organic substances. These manures are characterized by their bulk and provide (i) small amounts of plant nutrients and (ii) large quantities of organic matter. Among these, farmyard manure, compost, and green manure are the most significant and widely utilized. The impact of bulky organic manures on soils is three fold:

(i) Nutrient Supply: These manures, like commercial fertilizers, supply essential plant nutrients directly to plants. Although the nutrients are present in small quantities, large volumes of these manures are required per hectare. In addition to major nutrients, they also contain trace amounts of micronutrients.

(ii) Improvement of Soil Properties: Bulky organic manures enhance the organic matter content of soils, improving their physical properties. This is particularly beneficial for arable lands. These manures increase the humus content, temporarily improving the water retention capacity of sandy soils and enhancing drainage in clayey soils.

(iii) Support for Soil Microorganisms: By providing organic matter, bulky organic manures serve as food for soil microorganisms. This stimulates microbial activity, which helps in converting otherwise unavailable plant nutrients into forms that are accessible to plants.

1. Farmyard manure

Farmyard manure (F.Y.M.) is defined as the decomposed mixture of animal dung, urine, bedding material (litter), and leftover fodder or roughage fed to cattle. The manure collected daily from the cattleshed primarily consists of dung, with some urine absorbed by the refuse. Freshly collected F.Y.M. is in its raw form, whereas well-decomposed F.Y.M. has undergone a sufficient decomposition process over time. On average, fully decomposed farmyard manure contains 0.5% nitrogen (N), 0.2% phosphorus (P_2O_5), and 0.5% potassium (K_2O). For centuries, animal excreta has been used as manure, and farmyard manure is considered highly valuable by farmers and gardeners. As a result, F.Y.M. remains one of the most important agricultural by-products. However, it is unfortunate that nearly 50% of the cattle dung produced in India is used as fuel, thus being lost to agricultural purposes.

2. Compost from Farm and Town Refuse

Composting is primarily defined as a biological process in which organic matter is decomposed by both aerobic (requiring air or oxygen) and anaerobic (functioning without air or free oxygen) microorganisms, leading to a reduction in the carbon-to-nitrogen ratio of the material. The end product, known as compost, is produced as well-rotted manure. Due to its high organic matter content and relatively higher levels of major nutrients compared to farmyard manure, compost is considered valuable for enhancing and maintaining soil fertility in agricultural production. Vegetable and animal waste collected from farms or urban areas is used to produce compost. Therefore, the composition and nature of compost can vary. In India, the following types of compost are commonly produced:

(i) Compost from Farm Litter (Rural Compost): This type utilizes materials such as weeds, crop stubble, straw, crop residues (e.g., cotton stalks, groundnut husks), leaves, sugarcane trash, urine-soaked earth, cattle shed litter, waste fodder, and hedge clippings.

(ii) Compost from Town Refuse (Urban Compost): The primary components of urban compost include nightsoil, street refuse, household waste (commonly known as kachara), and wood ashes.

It is estimated that the major compost sources in India, including sewage, rural waste, and urban refuse, could potentially mobilize 4.064, 1.166, and 2.602 million tonnes of nitrogen (N), phosphorus (P_2O_5), and potassium (K_2O) annually. If 50% of these materials are utilized, this could result in a maximum savings of 2.032 million tonnes of N, 0.583 million tonnes of P_2O_5, and 1.301 million tonnes of K_2O. However, there are numerous challenges and social biases associated with the collection and processing of these wastes. To address this, administrators and social workers must develop effective methods for the collection and utilization of all waste materials.

(i) Compost from Farm Litter

In farms where cattle are not kept or where the cultivation is carried out in villages, farm litter such as weeds, stubble, bhusa, sugarcane trash (pachat), and other similar materials are converted into compost on the farm itself.

Method of Composting: The refuse collected from the farm is stored until a sufficient mass is accumulated for composting. A trench of appropriate dimensions, typically 457–610 cm long, 152–183 cm wide, and 91–183 cm deep, is dug. The collected refuse is mixed thoroughly and spread in layers, with each layer being about one foot thick. After moistening the first layer with a slurry of cow dung and water or a mixture of earth and water, a second layer of mixed refuse is added, continuing until the heap reaches a height of 46–61 cm above the ground level. The top is then covered with a thin layer of earth. After three months of decomposition, the material is taken out of the trench, shaped into a conical heap above the ground, moistened if needed, and covered with earth. After an additional one to two months, the compost becomes ready for application in agricultural fields.

Composition of Farm Litter Compost: The composition of farm litter compost can vary depending on the type of refuse used. However, it typically contains the following major plant nutrients:

Nitrogen (N): 0.5%

Phosphorus (P_2O_5): 0.15%

Potassium (K_2O): 0.5%

Sugarcane trash is one of the abundant types of farm refuse commonly used for composting.

Composting of Sugarcane Trash: There are two recommended methods for composting sugarcane trash: the "pit method" (anaerobic process) and the "heap method" (aerobic process).

- Pit Method: In the pit method, a pit of dimensions 183 cm wide, 91 cm deep, and 915 cm long (or any suitable length) is dug in an elevated area with minimal risk of water percolation. The materials for composting, such as sugarcane trash, cane leaves, cattle dung, and urine, are added in layers. Initially, a layer of sugarcane trash is spread, followed by a layer of cattle dung and urine. Optionally, ammonium sulphate can be added to accelerate decomposition. The entire mixture is layered until the height reaches 168–183 cm. During the summer months, one or two waterings may be necessary. Since the decomposition is anaerobic, the process is slower, and ammonia loss is minimal due to the stable formation of ammonium carbonate in the high-carbon dioxide environment. The compost is ready within four to six months, yielding six to eight tons per pit. A major advantage of the pit method is that it requires little to no additional attention after initial setup.

- Heap Method: In the heap method, a level portion of land near a water source is selected for the compost heap. The heap is made 6 to 8 feet wide, 107 to 122 cm high, and of a length based on the available material. The refuse is layered, with four to five layers of sugarcane trash stacked one over the other until the heap reaches a height of 76–122 cm. The top is covered with mud. After one and a half to two months, when the heap starts to subside, the first turning is done, mixing the outer material with the inside contents. Typically, two to three turnings are required for complete decomposition. This process takes approximately 4.5 to 5 months, making the compost suitable for application after this period. The heap method is especially effective during the rainy season and winter. However, a disadvantage is the need for multiple turnings, requiring additional time and labor.

(ii) Compost from town refuse or waste

India's urban population, which currently stands at approximately 125 million, has grown significantly. As urbanization increases, a larger proportion of agricultural produce is directed towards urban areas to meet the food and industrial needs of the population. Urban waste, including night-soil, sewage, sludge, street refuse, dustbin waste, factory by-products, and textile waste, contains essential nutrients like nitrogen, phosphorus, and potassium. This urban refuse can be transformed into high-quality manure through the composting process, commonly referred to as the "Bangalore Process."

Bangalore Method of Composting Urban Waste: To compost urban waste, trenches of suitable dimensions are dug at a designated site for composting, which should ideally be located at least two furlongs away from the town's edge and not on the windward side, as winds typically blow from the west. The trench dimensions vary according to the town's population, with the following recommendations:

- Population Below 5,000: Length 475 cm, Breadth 152 cm, Depth 76 cm
- Population 5,000–10,000: Length 610 cm, Breadth 183 cm, Depth 91 cm
- Population 10,000–20,000: Length 762 cm, Breadth 183 cm, Depth 91 cm
- Population 20,000–50,000: Length 915 cm, Breadth 213 cm, Depth 91 cm
- Population Above 50,000: Length 915 cm, Breadth 244 cm, Depth 107 cm

The trenches are arranged in rows, with the longer sides aligned parallel and the shorter ends in a straight line, leaving 152 to 213 cm of space between adjacent trenches. The composting process begins by spreading a layer of street refuse, or *katchara*, approximately 15 cm thick at the bottom of the trench using long-handled rakes. A 5 cm layer of night-soil is then added on top of the *katchara*. If the night-soil is too solid to flow naturally, it can be spread with a long-handled spade. The process alternates between layers of *katchara* and night-soil. At the end of each day, the pile is covered with a 22 cm layer of *katchara*. If available, a 5 cm layer of loose earth can be added to the top of the heap at the end of each day to help control odors, conserve moisture, prevent nitrogen loss, and reduce fly nuisance. Once the trench is filled to ground level, it is covered with *katchara* and earth, ensuring the pile rises about 30 cm above ground level. During the rainy season, it is advisable to shape the top of the heap into a dome to prevent water accumulation in the trench. After 3 to 4 months, the compost is ready for field application. Once the composting process is complete, the trenches are emptied, and fresh *katchara* and night-soil are added in layers for the next cycle of composting.

Composition of Urban Compost: When the correct ratio of *katchara* and night-soil is maintained and adequate time is allowed for decomposition, the resulting compost is a well-decomposed, odor-free material with the following nutrient content:

- Nitrogen (% N): 1.4
- Phosphorus (% P_2O_5): 1.0
- Potassium (% K_2O): 1.4

In comparison to farmyard manure, town compost produced from *katchara* and night-soil has a higher nutrient content and thus greater fertilizer value.

Advantages of the Bangalore Method of Composting: The Bangalore method offers several benefits, including:

- Effective control of foul odors.
- Elimination of pathogenic organisms, making it a hygienic and sanitary process.
- Reduction in fly nuisance by destroying fly larvae.
- Suppression of weed seed viability.
- Prevention of nitrogen losses.
- Due to the effective management of odors and flies, composting sites can be located a few furlongs away from the town.

3. Night soil

Night-soil refers to human excrement, both solid and liquid. In China, this material has been used directly on land for over two thousand years, contributing significantly to soil fertility. In Japan, night-soil accounted for 15.5% of the total nitrogen, 5.1% of the phosphorus, and 11.8% of the total potassium applied to soils in 1951. The annual potential of night-soil from India's population of 634 million is estimated to provide 8.6 million tonnes of dry matter, containing approximately 0.412 million tonnes of nitrogen, 0.314 million tonnes of phosphoric acid, and 0.172 million tonnes of potash. Night-soil is more nutrient-dense in terms of nitrogen (N), phosphorus (P_2O_5), and potassium (K_2O) compared to farmyard manure or compost. On an oven-dried basis, its average chemical composition is as follows:

- Nitrogen (% N): 5.5
- Phosphorus (% P_2O_5): 2.0
- Potassium (% K_2O): 2.0

In India, night-soil is occasionally applied directly to soil in trenches or pits, typically measuring 305–366 cm in length, 71–91 cm in width, and 22–28 cm in depth. The night-soil is deposited in these pits and covered with a layer of earth or *katchara*. This method is known as the "Poudrette System," with the dried material from these pits being referred to as poudrette. When night-soil is mixed with an equal volume of ash and 10% powdered charcoal, it forms an odorless material containing 1.3% nitrogen, 2.8% phosphoric acid, 4.1% potash, and 24.2% lime. Adding 40–50% sawdust to night-soil results in a dry, acidic poudrette that typically contains 2–3% nitrogen.

Drawbacks of the Poudrette System: The poudrette system has several disadvantages:

(i) It emits strong, unpleasant odors. During anaerobic decomposition of night-soil, various gases, such as hydrogen sulphide, are released.

(ii) It promotes fly breeding due to the conditions created during the decomposition process.

(iii) The anaerobic decomposition process results in significant nitrogen loss, as gases are released.

(iv) Since poudrette is only partially decomposed, it retains the offensive odor and other undesirable characteristics of night-soil, making it difficult to handle using regular labor. In some cases, workers from the *bhangi* class may be required for these tasks.

(v) The amount of manure produced by the poudrette system is relatively small compared to the volume of night-soil processed.

Improved Methods for Managing Night-Soil: Given that night-soil serves as a valuable organic manure, providing significant organic matter and essential plant nutrients to the soil, it is crucial to handle it using the following hygienic practices:

(i) Night-soil must be managed in a manner that prevents fly infestation and minimizes fly breeding.

(ii) Proper storage methods should be adopted to ensure that night-soil does not contaminate drinking water sources.

(iii) Pathogenic microorganisms, protozoan cysts, and parasitic worm eggs must be effectively eliminated before the night-soil is applied to agricultural land.

(iv) Efforts should be made to compost night-soil alongside other waste materials. This can be undertaken by municipal or town authorities in urban areas and by farmers themselves in rural regions.

4. Sewage and Sludge

In contemporary urban sanitation systems, human excreta and other wastes are removed using water, forming the basis of the sewage sanitation system. This process involves significant dilution, with water often comprising up to 99% of the total sewage volume. Sewage typically consists of two components:

Solid portion: Known as sludge.

Liquid portion: Referred to as sewage water.

Both components contain essential plant nutrients and can be used to enhance crop production. However, untreated sewage or sullage should not be applied to crops consumed raw, such as vegetables, due to the risk of bacterial contamination, which poses serious health concerns. To address these risks, both components of sewage undergo preliminary fermentation and oxidation treatments. These treatments reduce bacterial contamination, eliminate offensive odors, and narrow the carbon-to-nitrogen (C:N) ratio of the solid portion, making it safer and more suitable for agricultural use.

Sludges and Their Treatment

The solid portion of sewage, or sludge, is separated and treated before its use as manure. This involves allowing sewage to settle in tanks, which removes heavier solid matter and facilitates preliminary fermentation and oxidation of organic material. This process reduces the C:N ratio, transforming the sludge into "activated sludge," which is odorless and rich in nutrients. On a dry weight basis, activated sludge contains approximately 3–6% nitrogen, 2% phosphorus, and 1% potassium, making it a valuable fertilizer. Different methods of sewage treatment produce various types of sludge, classified as follows:

(i) Settled sludge: Formed through plain sedimentation.
(ii) Digested sludge: Produced by anaerobic decomposition of sedimented sludge.
(iii) Activated sludge: Generated by rapid aerobic treatment, which causes coagulation and settling of suspended materials.
(iv) Digested activated sludge: Further processed activated sludge.
(v) Chemically precipitated sludge: Created through chemical treatment to separate solids.

Sludge Production and Characteristics: Sludges are produced using various methods mentioned earlier, and their nutrient content can vary significantly depending on the composition of the sewage and the treatment process employed. On a dry-matter basis, the nutrient composition of sludges typically falls within the following ranges:

Nitrogen: 1.5–3.5%

Phosphorus: 0.78–4.0%

Potassium: 0.3–0.6%

In general, sludges are nutrient-rich in nitrogen and phosphorus but relatively low in potassium. The primary agronomic value of sludges lies in their slow-release nitrogen and phosphorus content, which can support sustained crop growth. Given the diverse range of materials entering sewage and the transportation of food into urban areas from multiple agricultural regions, sludges also serve as a potential source of micronutrients, including boron, manganese, copper, zinc, and iron.

Sewage Irrigation: When raw sewage undergoes treatment to separate the solid portion (sludge), the resulting liquid, known as treated effluent, can be utilized for irrigation. This practice, referred to as sewage irrigation, provides field crops with essential nutrients such as nitrogen, phosphorus, and potassium. Effluents from settling tanks that have undergone only anaerobic fermentation still contain considerable colloidal matter, which may be objectionable. However, further aerobic oxidation transforms the effluent into a clear, odorless liquid containing nitrates and significantly reduces the number of pathogenic bacteria originally present. Both activated sludge and treated effluent can be safely applied to most field crops. However, they should not be used on vegetables consumed raw or uncooked to prevent health risks. In India, approximately 32 municipal committees manage the utilization of around 82,000 cubic meters of partially treated sewage water daily, irrigating about 17,400 hectares of agricultural land. The application rate and crop rotation practices are carefully monitored to suit soil conditions and prevent the occurrence of sewage-related soil issues, commonly referred to as sewage sickness. The use of treated sewage and sludge represents an effective method for recycling urban organic waste into agriculture, enhancing soil fertility while minimizing waste disposal challenges.

Crops Commonly Cultivated with Sewage Irrigation: Field crops that exhibit rapid vegetative growth are typically cultivated using sewage irrigation. These crops include:

- Fodder crops: Examples include oats, sorghum (jowar), maize, berseem, and lucerne.
- Sugarcane: A widely cultivated crop for its high water demand and nutrient requirements
- Vegetable crops: Examples include cabbage, cauliflower, turnip, potato, brinjal (eggplant), okra, beans, and leafy vegetables.

While sewage water has been utilized for irrigation over many years, its application is not universally recommended for all crops. Specific state regulations and medical guidelines are in place to ensure safe practices. In particular, vegetable crops whose marketable produce is consumed raw, such as tomatoes, radish, onion, garlic, and carrot, should not be irrigated with sewage water. This precaution is necessary to mitigate the risk of disease transmission and protect public health.

B. Concentrated organic manures

Concentrated organic manures are organic materials characterized by higher concentrations of essential plant nutrients, such as nitrogen, phosphorus, and potassium, compared to bulky organic manures like farmyard manure and compost. These manures are derived from raw materials of either animal or plant origin. Common examples include oilcakes, blood meal, fish manure, meat meal, and wool waste.

1. Oilcakes

India produces a wide variety of oilcakes, which can be broadly categorized into two types:

Edible oilcakes: Suitable for feeding livestock.

Non-edible oilcakes: Unsuitable for livestock consumption.

Although edible oilcakes are primarily intended as cattle feed, they are sometimes applied to the soil as organic manure. Both types of oilcakes are utilized as concentrated organic manures. Annually, approximately two million tons of oilcakes are produced in India. Oilcakes not only supply nitrogen but also contribute phosphorus, potassium, and a significant amount of organic matter to the soil. The chemical composition of commonly available oilcakes in India varies, with nitrogen content ranging from 2.5% in mahua cake to 7.9% in decorticated safflower cake (Table 1). They also contain phosphorus (0.8–2.9% P_2O_5) and potassium (1.2–2.2% K_2O) in smaller amounts. Due to their high nitrogen content, oilcakes are classified as organic nitrogenous manures. Oilcakes act as quick-releasing organic manures. Despite being insoluble in water, the nitrogen they contain becomes available to plants within 7–10 days of application. An exception to this is mahua cake, whose nitrogen release takes approximately two months. Consequently, mahua cake should be applied two months before sowing, provided the soil retains adequate moisture. This oilcake is particularly suitable for perennial crops and long-duration crops such as sugarcane and tapioca.

Table 1: Average nutrient contents of principal oilcakes

Name of the oilcake	N (%)	P_2O_5 (%)	K_2O (%)
Non-edible oilcakes			
Castor cake	4.3	1.8	1.3
Cottonseed cake undercorticated	3.9	1.8	1.6
Karanj or honge cake	3.9	0.9	1.2

Name of the oilcake	N (%)	P_2O_5 (%)	K_2O (%)
Mahua or ippi cake	2.5	0.8	1.8
Neem cake	5.2	1.0	1.4
Safflower cake undecorticated	4.9	1.4	1.2
Undi or punna cake	3.6	1.5	2.0
Edible oilcakes			
Coconut cake	3.0	1.9	1.8
Coconut seed cake decorticated	6.4	2.9	2.2
Groundnut cake	7.3	1.5	1.3
Linseed cake	4.9	1.4	1.3
Jambo cake	4.9	1.6	1.9
Niger cake	4.7	1.8	1.3
Rapeseed cake	5.2	1.8	1.2
Safflower cake decorticated	7.9	2.2	1.9
Sesame or til cake	6.2	2.0	1.2

Oilcakes should be finely powdered before application to ensure even distribution and facilitate decomposition by soil microorganisms. They can be applied either a few days before sowing or at the time of sowing. Groundnut cake, in particular, is widely used as a top-dressing for sugarcane. Depending on the crop and agronomic practices, oilcakes are applied through broadcasting, drilling, or placement near the root zone during earthing up. Oilcakes produced using traditional methods, such as country *ghani* or *gana*, generally retain a slightly higher oil content compared to those processed by hydraulic or expeller presses. Due to their elevated oil content, these oilcakes are slower to act, as the oil inhibits the rapid conversion of organic nitrogen into plant-available forms.

2. Bloodmeal

Dried blood, commonly referred to as bloodmeal, typically contains 10–12% nitrogen and 1–2% phosphoric acid. On average, an adult cow yields approximately 13.6 kg of blood, while a goat or sheep provides around 1.36 kg. However, the current practices in slaughter houses often fail to adequately collect blood. To address this, slaughter houses should be equipped with durable concrete floors featuring central drainage systems that direct blood into storage tanks. The preparation of bloodmeal begins by treating the blood clots with commercial copper sulphate at a rate of 125 g per 100 kg of clot. The clots are then evaporated to dryness using a sand bath, spread on a concrete surface covered with a net, and sun-dried. Once fully dried, the material is ground into powder, packaged, and marketed as bloodmeal. Producing one tonne of bloodmeal costs approximately 135 rupees. This cost can be further reduced when processing large quantities of clotted blood, making bloodmeal production economically viable in facilities that slaughter around 140 sheep or goats, or 10 adult cattle, daily. Bloodmeal is a fast-acting organic manure suitable for all soil types and crops. Its application method is similar to that of oilcakes.

3. Meatmeal

Meatmeal contains about 10.5% nitrogen and 2.5% phosphoric acid. An adult animal typically yields 36–45 kg of meat post-slaughter. If all the meat were processed into meatmeal, India could potentially produce around 100,000 tonnes of this organic manure annually. Currently, meatmeal production is limited and sold under various trade names. The manufacturing process for meatmeal involves cooking or digesting the bones and meat in specialized pans for 2–3 hours. The bones are then separated from the meat, which is dried on a sand bath until it becomes brittle, after which it is ground into powder. Alternatively, double-jacket trays powered by steam can be used to dry the material within an hour. A device known as a mutton squeezer can also extract water from the meat before final drying on a pan. Meatmeal acts as a rapid-release organic fertilizer and is effective across all soil types and crops. It is applied in a manner similar to oilcakes.

4. Fish Manure

Non-edible fish carcasses and offal are commonly utilized to produce fishmeal, an organic fertilizer. The preparation process involves drying the fish, crushing or powdering it, and then packaging it into bags. Specialized equipment such as fish boilers and squeezers can be employed to expedite the processing. Fish manure, available in forms such as dried fish, fishmeal, or fish powder, typically contains 4–10% nitrogen, 3–9% phosphorus pentoxide (P_2O_5), and 0.3–1.5% potassium oxide (K_2O). The nutrient composition varies depending on the type of fish used. Fishmeal is a fast-acting organic fertilizer suitable for application to all types of crops and soils. It is recommended to powder the fishmeal before application to enhance its effectiveness.

5. Horn-and-Hoof Meal

An adult livestock animal yields approximately 6–8 pounds of horn and hoof material. These components, after undergoing cooking in a bone digester, are removed, dried, and ground into a fine powder. Horn-and-hoof meal contains around 13% nitrogen, making it a rich source of this nutrient. India has the potential to produce approximately 14,000 tonnes of horn-and-hoof meal annually, provided dead animals are systematically collected and processed. This organic manure is a valuable resource for enhancing soil fertility and crop productivity.

C. Greem manuring

Green manuring is the agricultural practice of incorporating undecomposed green plant material into the soil to enhance its physical structure and fertility. This age-old method of ploughing green crops into the soil has been widely practiced to improve soil conditions.

Types of Green Manuring

1. *Green Manuring In Situ:* In this method, green manure crops are cultivated and subsequently incorporated into the same field intended for green manuring. These crops may be grown as pure stands or intercropped with the main crop. Commonly used green manure crops in this system include sunnhemp (*Crotalaria juncea*), dhaincha (*Sesbania aculeata*), pillipesara (*Phaseolus trilobus*), and guar (*Cyamopsis tetragonoloba*).
2. *Green-Leaf Manuring*: This involves incorporating green leaves and tender twigs collected from shrubs and trees grown on field bunds, wastelands, or nearby forested areas into the soil. Common plant species used for green-leaf manuring include Glyricidia (*Glyricidia maculata*), *Sesbania speciosa*, and karanj (*Pongamia pinnata*). While green manuring in situ is more prevalent in northern India, green-leaf manuring is widely practiced in eastern and central India.

Advantages of Green Manuring

Green manuring offers several benefits for soil fertility and agricultural productivity:

1. *Increased Organic Matter*: Incorporating green plant material into the soil boosts organic matter, stimulating the activity of soil microorganisms.
2. *Nutrient Recycling*: Nutrients absorbed by the green manure crops from deeper soil layers are returned to the upper soil profile, enhancing nutrient availability.
3. *Improved Soil Structure*: Green manuring enhances soil structure, promoting better aeration and aggregation.
4. *Enhanced Water Infiltration*: It facilitates the penetration of rainwater, reducing runoff and soil erosion.
5. *Nutrient Retention:* Green manure crops help prevent nutrient losses caused by leaching.
6. *Nitrogen Fixation*: When leguminous crops such as *sunnhemp* and *dhaincha* are used, they fix atmospheric nitrogen, enriching the soil for subsequent crops.
7. *Improved Nutrient Availability*: Green manuring increases the availability of essential nutrients such as phosphorus (P_2O_5), calcium, potassium, magnesium, and iron in the soil.

Disadvantages of Green Manuring

While green manuring offers several benefits, certain limitations may arise if the practice is not executed properly or if environmental conditions are unfavorable. The key disadvantages include the following:

1. Dependence on Rainfall in Rainfed Areas: In rainfed regions, inadequate rainfall after incorporating the green manure crop into the soil can hinder its proper decomposition and negatively impact the germination of the subsequent crop. This challenge is particularly significant in wheat-growing areas of India.
2. Economic Constraints in Irrigated Regions: In areas with irrigation facilities, green manuring may result in the loss of the *Kharif* crop, making it economically unviable. Farmers in these regions may prefer to grow *Kharif* crops or rely on readily available chemical fertilizers instead of green manuring.
3. Cost Considerations for Nitrogen Addition: If the primary goal of green manuring is nitrogen enrichment, the expense of cultivating green manure crops may exceed the cost of purchasing chemical fertilizers with equivalent nitrogen content.
4. Increased Pest and Disease Risks: Green manuring may contribute to a rise in certain pests, diseases, and nematode populations, which could harm subsequent crops.
5. Uncertainty in Crop Establishment: Achieving a good stand and optimal growth of green manure crops can be challenging in areas with insufficient or erratic rainfall, increasing the risk of failure.

14

Classification of Chemical Fertilizers

- Just as humans require proper nutrition to remain healthy, plants also need specific nutrients for proper growth.
- Fertilizers are substances applied to soil to enhance plant growth and productivity. They can be natural or synthetic materials added to the soil or land to boost fertility.
- Fertilizers act as nutritional supplements for plants, providing the 16 essential nutrients they need to thrive.

Fertilizer: A fertilizer is a substance, either natural or synthetic, that is added to soil or plant tissues to provide essential nutrients required for plant growth.

Complete v/s Incomplete fertilizer

A *complete fertilizer* contains all three primary nutrients: nitrogen, phosphorus, and potassium. Examples include formulations like 10-10-10, 15-30-15, and 20-5-20.

An *incomplete fertilizer* lacks one or more of the primary nutrients. Examples include 20-0-0, 0-20-0, and 12-0-44%.

A *straight fertilizer* provides only a single primary nutrient, such as urea or ammonium sulfate.

Mixed or complex fertilizers contain two or more nutrient elements.

Basic Forms of Fertilizers

- Granular fertilizers are formulated to be incorporated into the soil and are water-soluble.
- Powdered fertilizers are intended to be dissolved in water for application as a liquid feed to both soil and foliage.
- Liquid fertilizers are typically concentrated solutions that require dilution before use.

Classification of fertilizers

1. Straight fertilizers provide a single primary plant nutrient, such as nitrogen, phosphorus, or potassium. Examples include urea, ammonium sulfate, and potassium chloride.

2. Complex fertilizers contain two or three primary plant nutrients, with at least two of them chemically combined. They are commonly produced in granular form. Examples include diammonium phosphate, nitrophosphates, and ammonium phosphate.
3. Mixed fertilizers are physical blends of straight fertilizers, containing two or three primary plant nutrients. They are prepared by thoroughly mixing the components, often using mechanical methods.

1. Based on concentration of primary plant nutrients (N, P and K fertilizers)

- Low analysis fertilizers: These fertilizers contain less than 25% of primary nutrients. Example: SSP (16% P_2O_5).
- High analysis fertilizers: The total content of primary nutrients in these fertilizers is above 25. Example: Urea (46% N), Anhydrous ammonia (82.2% N), Ammonium phosphate (20% N + 20% P_2O_5).

2. Based on physical form

- Solid fertilizers: They are of two types such as powder (single superphosphate) and crystals (Ammonium phosphate). Examples: prills (urea, diammonium phosphate, superphosphate), granules (holland granule), supergranules (urea supergranules) and briquettes (urea briquettes).
- Liquid fertilizers: They are applied with irrigation water or for direct application. They are of two types (i) Clear liquid fertilizers :When the fertilizers are completely dissolved in water and (ii) Suspension liquid fertilizers: When fertilizers are suspended as fine particles in water.

3. Based on nature of fertilizers

Acid forming fertilizers	**Alkaline forming fertilizers or Basic fertilizers**
Fertilisers leave a residue of an acid in the soil	Fertilisers leave a residue of alkaline in the soil.
Alkaline soils are applied with acid forming fertilisers.	Acid soils are applied with basic fertilizers.
The acid equivalent refers to the amount of calcium carbonate needed to neutralize the acid residue.	
Fertilizer (Acid equivalent) Ammonium chloride: 128; Ammonium sulphate: 110; Ammonium sulphate nitrate: 93; Ammonium phosphate: 86 and Urea: 80	*Fertilizer (Basicity equivalent)* Sodium nitrate: 29; Calcium nitrate: 21; Calcium cyanide: 63 and Dicalcium phosphate: 25

Fertilizer grade: Fertilizer grade indicates the minimum percentage of nitrogen (N), phosphorus (P_2O_5), and potash (K_2O) present in a fertilizer. Fertilizer analysis represents the weight of nitrogen, phosphorus, and potassium as a percentage.

Nitrogenous fertilizers

Nitrogenous fertilizers are chemical compounds that supply nitrogen, an essential nutrient for plants, in forms that can be readily absorbed, primarily as ammonium (NH_4^+) or nitrate (NO_3^-). Some fertilizers release these forms of nitrogen following chemical or biological conversion processes.

Classification of Mineral Nitrogen Fertilizers

Chemical formula of nitrogen	Fertilizer	Chemical formula	N content (%)
Ammonium (NH_4^+)	Gaseous ammonia	NH_3	82
	Ammonia water	NH_3, NH_4OH	24
	Ammonium sulphate	$(NH_4)_2SO_4$	21
Nitrate (NO_3^-)	Calcium nitrate	$Ca(NO_3)_2$	16
	Sodium nitrate	$NaNO_3$	16
	Ammonium sulphate nitrate	$(NH_4)_2SO_4+NH_4NO_3$	26
Ammonium nitrate (NH_4^+), (NO_3-)	Ammonium nitrate	NH_4NO_3	35
	Calcium ammonium nitrate	$NH_4NO_3+CaCO_3$	25
Amide (NH_2^-)	Urea	$CO(NH_2)_2$	46
	Calcium cyanamide	$CaCN_2$	22

Synthesis and Properties of Nitrogenous Fertilizers

Ammonia

Ammonia is synthesized at high temperatures (approximately 550°C) and pressures (200 atmospheres) by passing a mixture of nitrogen (N_2) and hydrogen (H_2) in a 1:3 molar ratio over an activated iron oxide catalyst. This process is conducted in a specialized ammonia synthesis reactor, typically a steel cylinder with a diameter of 80–140 cm and a height of 10–18 meters. The reactor contains a catalyst bed designed to efficiently dissipate the heat generated during the reaction.

Physical and Chemical Properties of Ammonia

- Ammonia is a colorless gas with a sharp, pungent odor. In high concentrations, it is toxic to both humans and plants, but in diluted forms, it serves as a harmless and essential nutrient for plant growth.
- Ammonia contains 82% nitrogen by weight, making it a highly concentrated source of nitrogen for fertilizers. Approximately 99.5% of nitrogen-based fertilizers are produced using ammonia, as it is the most cost-effective nitrogen fertilizer to manufacture.
- At 10°C, ammonia gas exerts a pressure of 34.10 kg/in², while at 38°C, the pressure increases to 89.50 kg/in². It is typically stored in steel or plastic tanks capable of withstanding pressures up to 120.4 kg/in².

- Ammonia is transported as a pressurized liquid and released into the soil as a gas through ammonia-resistant hoses. It is reactive with metals such as copper and brass, so these materials should not be used for handling ammonia.
- Ammonia is water-soluble up to 30% by weight, forming a low-pressure solution known as aqua ammonia, which contains 24% nitrogen.
- Chemically, ammonia is a weak alkali.

Aqua Ammonia (Aqueous Ammonia)

Aqua ammonia, also known as aqueous ammonia, is produced by dissolving anhydrous ammonia in water. The process involves controlling the flow rates of both water and ammonia while managing the heat released during absorption through cooling mechanisms. Additionally, the concentration of the resulting aqua ammonia is carefully monitored during production.

Physical and Chemical Properties of Aqua Ammonia

- Aqua ammonia is a colorless solution with a characteristic pungent odor, typically containing 20% nitrogen.
- At 20°C, one liter of water can dissolve up to 700 liters of ammonia gas.
- It is a non-pressurized solution and does not contain free ammonia in its gaseous form.
- Aqua ammonia can bc stored and handled without the need for high-pressure tanks or specialized equipment.

Ammonium Sulphate [$(NH_4)_2SO_4$]

Ammonium sulphate is one of the oldest synthetic fertilizers, primarily linked to the steel manufacturing process, which consumes coke. Its production involves several methods, as detailed below:

Manufacturing Processes

1. By-Product Process
2. Gypsum process
3. Neutralization process

1. By-Product Process (Raw Materials: Coal and sulfuric acid)

During the production of iron and steel, bituminous coal undergoes destructive distillation (heating at 982.2°C in the absence of air), yielding coke and coke oven gas. This gas contains approximately 10% ammonia by volume, along with carbon monoxide, hydrogen, methane, water vapor, tar particles, and hydrocarbons. Ammonia is extracted by cooling the gas and bubbling it through water to form ammonia liquor, which is distilled and then reacted with dilute correc acid in saturators to form ammonium sulphate crystals. The crystals are then centrifuged, washed, and dried.

$2NH_3$ (g) + H_2SO_4 → $(NH_4)_2SO_4$ (salt) Energy 67.71kCal/kg

2. Neutralization Process

Raw Materials: Ammonia (from the Haber-Bosch process) and sulfuric acid

In this method, gaseous ammonia is directly neutralized with liquid sulfuric acid to produce ammonium sulphate. This exothermic reaction releases significant energy.

$$2NH_3\ (g) + H_2SO_4\ (liq) \rightarrow (NH_4)_2SO_4\ (salt) + 67.71\ kCal/g/mole$$

3. Gypsum Process or Leuna Process (Raw materials: NH_3, CO_2 and Gypysum)

Raw Materials: Ammonia, carbon dioxide, and gypsum (calcium dihydrate, $CaSO_4.2H_2O$)

In this method, anhydrous ammonia is absorbed in water to form ammonium hydroxide. Carbon dioxide is introduced under pressure, forming ammonium carbonate, which reacts with gypsum suspended in water. This double decomposition reaction yields ammonium sulphate and calcium carbonate. The insoluble calcium carbonate is filtered out, and ammonium sulphate is crystallized by evaporation.

$$NH_3 + H_2O \rightleftharpoons NH_4OH + 8.32kCal/g\text{–}mole$$

$$2NH_4OH + CO_2 \rightarrow (NH_4)_2CO_3 + H_2O + 22.08kCal/g\text{–}mole$$

$$(NH_4)_2CO_3 + CaSO_4.2H_2O \rightleftharpoons (NH_4)_2SO_4 + CaCO_3 + 2H_2O + 3.9kCal/g\text{–}mole$$

Properties of Ammonium Sulphate

Physical Properties

- Ammonium sulphate is a white crystalline salt, although the commercial product may vary in color from light yellow to grey and is free-flowing.
- The presence of thiocyanates in the product can be toxic to plants.
- It is stable in handling and storage, though it may cake if fine powders are present (typically stored in polythene bags).
- Bulk Density: 876.60 kg/m^3

Chemical Properties

- Ammonium sulphate contains 20.6% nitrogen and 23.45% sulphur, making it an acid-forming fertilizer.
- It is highly soluble in water at ordinary temperatures
- Solubility: 70.6 g/100 g water at 0°C and 103.8 g/100 g water at 100°C.
- It has a free acidity of 0.025% by weight (pH = 5.0).
- To neutralize the acidity produced by 1 kg of nitrogen from ammonium sulphate, 5.1 kg of lime is required.

Calcium Ammonium Nitrate (CAN)

Calcium ammonium nitrate (CAN) is one of the primary straight nitrogen fertilizers produced in India. It is also referred to as Lime Ammonium Nitrate.

Raw Materials: Ammonia, Nitric acid, Limestone or dolomite and Soapstone (magnesium silicate)

Manufacturing Process

The production of CAN begins with the neutralization of anhydrous ammonia and nitric acid, which are pre-heated to 85°C and 65°C, respectively, using steam vapor. This reaction produces ammonium nitrate liquor with a concentration of 82–83%. The solution is then further concentrated to 92–94% in a vacuum concentrator using steam and stored in a dedicated tank. $NH_3(g) + HNO_3\ (l) \rightarrow NH_4NO_3\ (l) + 26\,kcal/mol$

The concentrated ammonium nitrate solution (92–94%) is sprayed into a granulator along with a measured quantity of limestone powder. This process forms hot granules of calcium ammonium nitrate. The granules are then dried in a rotary drier using hot air and subsequently screened to achieve the desired size. The granules are cooled in a rotary cooler using air and coated with soapstone (talc) in a coating drum. The final product is a blend of ammonium nitrate and calcium carbonate.

$NH_4NO_3 + CaCO_3 \rightarrow NH_4NO_3.CaCO_3$

Physical Properties

- CAN is a free-flowing granular material with a particle size ranging from 1 to 4 mm.
- The addition of calcium carbonate (lime powder) enhances the handling properties of CAN and minimizes the explosive and hygroscopic characteristics of ammonium nitrate.

Chemical Properties

- CAN contains 25% nitrogen, evenly distributed between ammonium (NH_4^+) and nitrate (NO_3^-) forms, each contributing 12.5%.
- It is a neutral fertilizer, leaving neither acidic nor basic residues upon application to soil.
- It contains 8.1% calcium and 0.5% calcium nitrate by weight.
- CAN is highly soluble in water.

Urea (Carbamide, NH_2CONH_2)

Urea, also known as carbamide, is a "non-ionic" nitrogenous compound widely used as a fertilizer for crops and as a protein supplement in ruminant feed. It is the most significant nitrogen fertilizer, accounting for approximately 91% of

India's total nitrogen fertilizer production. Its widespread use is primarily due to its exceptionally high nitrogen content (46%).

Raw Materials: Ammonia and carbon dioxide

All industrial processes for urea production rely on the dehydration of ammonium carbamate. The chemical reaction is represented as:

$2NH_3+CO_2 \rightarrow 2500-3500\,psi\,180-200^{\circ}C\,NH_4COONH_2$ (Ammonium carbamate)

$$2NH_3 + CO_2 \xrightarrow[\text{2500-3500 psi}]{\text{180-200 °C}} \underset{\text{Manufacturing}}{NH_4CO_2NH_2} \longrightarrow \underset{\text{Urea}}{NH_4CONH_2} + H_2O$$

Liquid ammonia and carbon dioxide gas are continuously introduced into a reactor operating at a temperature of 180–200°C and a pressure of 2500–3500 psi. This exothermic reaction yields a mixture containing urea, ammonia, carbon dioxide, ammonium carbamate, and water. The mixture is then passed into carbamate strippers, where the liquid phase (containing urea, water, and small amounts of ammonium carbamate and ammonia) is separated from the gaseous phase (containing ammonia, carbon dioxide, and water vapor). The resulting aqueous urea solution contains approximately 70–80% urea. While this solution can be used directly in nitrogen solutions, further concentration is required to produce solid urea.

Biuret Formation

When the concentrated urea solution is exposed to temperatures exceeding 100°C during the prilling process, biuret is formed due to the interaction of two urea molecules. Biuret is toxic to plants and forms as follows:

$$\underset{\text{Urea}}{NH_2CO_2NH_2 + NH_2CO_2NH_2} \longrightarrow \underset{\text{Biuret}}{2NH_2—CO—NH—CO—NH_2} + NH_3$$

The urea solution is concentrated to 99.7–99.8% purity using vacuum evaporation and then solidified by spraying into a prilling tower, forming prills or granules.

Physical Properties

- Urea is a white organic compound with low bulk density (~0.7 kg/L).
- It is a solid fertilizer, commonly available as granules (1–2 mm in size).
- It has a specific gravity of 1.335.

Chemical Properties

- Urea is a synthetic organic compound and the richest nitrogen source (46% nitrogen in amide form) among solid nitrogen fertilizers.
- It is highly soluble in water, with a solubility of 100 g per 100 g of water at 20°C.

- The biuret content is approximately 1.5% by weight.
- Its chemical composition is identical to the urea naturally found in animal urine.

15

Nanofertilizers

Definition and Importance

In recent years, nano fertilizers have emerged as an innovative solution to maximize both the quantity and quality of agricultural production. These fertilizers are synthesized or modified versions of conventional fertilizers, bulk materials, or plant-derived extracts obtained through various chemical, physical, mechanical, or biological methods leveraging nanotechnology. Nano fertilizers play a critical role in enhancing soil fertility, crop productivity, and the quality of agricultural produce. The defining characteristics of nano fertilizers lies in their extremely small particle size, typically less than 100 nanometers (nm), which results in a significantly higher surface area. This increased surface area provides more active sites for facilitating various metabolic processes within the plant system, thereby boosting photosynthate production. Their small size and high reactivity enhance their solubility in water and other solvents, improving nutrient uptake and utilization when applied to soil or leaves.

Advantages of Nano Fertilizers Over Traditional Fertilizers

1. *Improved Nutrient Use Efficiency (NUE):* Nano fertilizers enhance NUE, reduce the excessive use of chemical fertilizers, and improve soil fertility and crop yield. Their small size allows frequent application in smaller quantities, making them environmentally sustainable.
2. *Enhanced Crop Productivity and Quality:* Nano fertilizers positively impact crop yield and quality parameters. They are non-toxic, less harmful to the environment and humans, and help minimize input costs while maximizing profits. Their ability to improve the nutritional content and taste of crops, such as increasing protein levels in wheat grains, further highlights their significance.
3. *Environmental and Economic Benefits:* By increasing NUE, nano fertilizers reduce the environmental cost of excessive fertilizer use. They also promote balanced crop fertilization, enhancing plant growth by improving disease resistance, stabilizing plants against bending, and fostering deeper rooting systems.
4. *Seed Germination and Seedling Growth:* Research indicates that nano fertilizers significantly enhance seed germination and seedling development

by improving nutrient availability to the growing plant. This leads to healthier seedlings with increased shoot and root lengths. However, excessive concentrations may have inhibitory effects.

Specific Applications and Observations

- Nano zinc oxide (ZnO) has been shown to improve seed germination and root growth in peanuts compared to bulk zinc sulphate.
- Nano-scale silicon dioxide (SiO_2) positively affects germination in soybean.
- Nano fertilizers promote chlorophyll formation, increase photosynthesis rates, enhance dry matter accumulation, and support overall plant growth.

Nano fertilizers represent a transformative advancement in agricultural practices, providing an effective, sustainable, and environmentally friendly alternative to conventional fertilizers. With their ability to enhance crop yield, quality, and resource efficiency, they hold significant potential to address global challenges in agriculture while promoting sustainable development.

16

Fertilizer Control Order and Integrated Nutrient Management

Fertilizer Control Order (FCO) and Essential Commodities Act, 1955

The Fertilizer Control Order (FCO) was introduced under the Essential Commodities Act, 1955, and came into effect in 1957. Its primary purpose was to regulate the sale, price, and quality of fertilizers to ensure their availability and effectiveness for agricultural use. Over the years, approximately 70 amendments were made to address various issues, such as ensuring the quality of fertilizers and enforcing quality control measures. These changes culminated in the updated Fertilizer Control Order, 1985, which came into force on September 25, 1985.

Fertilizer Control Order, 1985

The FCO, 1985, provides a comprehensive framework for the regulation of fertilizers, including their specifications, quality standards, sampling methods, and certification. It is divided into the following key components:

a) Schedule I

Part A - Specifications of Fertilizers

This section outlines the specific chemical and physical properties required for different fertilizers, ensuring they meet the standards necessary for effective use in agriculture.

Part B - Tolerance Limits in Plant Nutrients for Various Fertilizers

This section specifies the permissible limits of variation in the nutrient content of fertilizers. These tolerance limits ensure uniformity in quality and prevent malpractices in the fertilizer industry.

b) Schedule II

Part A - Procedure for Drawing Samples of Fertilizers

This section details the protocols for collecting representative samples of fertilizers to verify their compliance with the specifications outlined in Schedule I.

Part B - Methods of Analysis of Fertilizer

This section describes the analytical procedures to assess the chemical and physical properties of fertilizers, ensuring their adherence to quality standards.

c) Forms of Registration

This section provides the formats and procedures for manufacturers, distributors, and retailers to register under the FCO for the production, distribution, and sale of fertilizers.

d) Appendices

The appendices cover a range of additional provisions and guidelines, including:

- Sale of Small Quantities of Fertilizers: Provisions for selling fertilizers in smaller packages to meet the needs of small-scale farmers.
- Marking on Containers: Guidelines for labeling fertilizer containers with essential information such as nutrient content, manufacturer details, and batch numbers to ensure traceability and transparency.
- Equipment Required for Laboratory: Specifications for the equipment necessary in laboratories to conduct quality control tests on fertilizers.
- Forms for Obtaining Certification: Standardized forms and procedures required for obtaining certification to manufacture or sell fertilizers under the FCO.

Significance of the Fertilizer Control Order, 1985

The FCO plays a critical role in ensuring that fertilizers meet prescribed quality standards, which is essential for maintaining soil fertility, improving crop productivity, and preventing adulteration or misbranding. By regulating the production, distribution, and sale of fertilizers, the FCO safeguards the interests of farmers and promotes sustainable agricultural practices.

Integrated Nutrient Management (INM)

Integrated Nutrient Management (INM) is a holistic approach that combines and optimizes the use of multiple nutrient sources, such as chemical fertilizers, organic amendments, and biological inputs, to meet the nutrient requirements of crops sustainably. This approach emphasizes the judicious and balanced application of nutrients without causing degradation to soil, water, or the environment.

Key Principles and Objectives of INM

1. Efficient Nutrient Utilization: INM integrates various nutrient sources to achieve an optimal nutrient supply for crops while minimizing losses to the environment.
2. Sustainable Soil Health: It aims to enhance soil fertility and productivity by improving the physical, chemical, and biological properties of the soil.

3. Prevention of Nutrient Deficiencies: INM addresses macronutrient and micronutrient deficiencies in crops, ensuring better plant growth and higher yields.
4. Environmental Protection: By reducing nutrient losses through leaching, denitrification, and volatilization, INM mitigates the environmental impacts of agriculture, such as groundwater contamination and greenhouse gas emissions.

Components of Integrated Nutrient Management

1. Chemical Fertilizers: Chemical fertilizers remain a central component of INM, providing readily available nutrients to crops. However, their use is carefully calibrated to prevent over-application, which could harm soil health and the environment.
2. Organic Amendments: Organic inputs, including animal manures, composts, press mud, crop residues, rural and urban wastes, and green manure, play a vital role in INM. These inputs improve soil structure, increase microbial activity, and enhance nutrient retention.
   - Green Manuring: Particularly relevant in rice cultivation, green manures like leguminous crops (e.g., *Crotalaria* and *Sesbania*) contribute to nitrogen fixation. While their N-substitution potential is significant, they have a limited residual value compared to other organic manures.
   - Residual Effects of Organic Manures: Organic manures not only supply nutrients immediately but also leave a residual impact, improving soil fertility over time.
3. Biological Inputs: Biological nitrogen fixation (BNF) is an integral part of INM, achieved through:
   - Symbiotic Nitrogen Fixation: Rhizobia in legume root nodules fix atmospheric nitrogen into forms usable by plants.
   - Non-Symbiotic Nitrogen Fixation: Free-living nitrogen-fixing bacteria (e.g., *Azotobacter*, *Azospirillum*) and cyanobacteria in the soil contribute to nitrogen enrichment.

Benefits of Integrated Nutrient Management

1. Enhanced Crop Yields: Combining organic and inorganic nutrient sources ensures a balanced nutrient supply, leading to sustained and higher crop productivity.
2. Improved Soil Health
   - Organic inputs contribute to soil organic matter buildup, enhancing soil physical properties, such as porosity and water-holding capacity.
   - Biological activity in soil increases due to organic inputs, promoting nutrient cycling and improving soil fertility.

- Carbon sequestration from organic amendments helps in mitigating climate change while improving soil structure.

3. Reduced Nutrient Losses
    - The integration of organic sources with chemical fertilizers reduces losses of nitrogen through denitrification and leaching, enhancing nitrogen use efficiency (NUE).
    - Controlled release fertilizers and split applications minimize environmental losses.
4. Environmental Sustainability: INM reduces dependency on chemical fertilizers, thereby minimizing the risk of water pollution, eutrophication, and greenhouse gas emissions.

Challenges in Integrated Nutrient Management

1. Availability of Organic Resources: Limited availability of quality organic inputs such as composts, green manures, and livestock wastes can constrain the adoption of INM practices.
2. Labor and Time Requirements: The preparation and application of organic inputs, such as composts and green manures, are labor-intensive and time-consuming.
3. Knowledge and Training: Farmers may lack adequate knowledge of the nutrient value of organic and biological inputs, as well as proper techniques for their integration with chemical fertilizers.
4. Economic Considerations: In certain cases, the cost of implementing INM may exceed that of using only chemical fertilizers, especially in regions where organic inputs are scarce or expensive.

Integrated Nutrient Management is a sustainable and eco-friendly approach to maintaining soil fertility and crop productivity. By combining chemical fertilizers, organic amendments, and biological inputs, INM ensures a balanced nutrient supply while preserving soil health and minimizing environmental impacts. Its successful implementation requires awareness, training, and support for farmers, along with policies promoting the availability and use of organic and biological inputs.

17

Soil Amendments

Soil Amendments and Their Role in Enhancing Soil Health

Soil amendments are substances that positively influence plant growth by inducing one or more beneficial effects in the soil. These effects include:

1. Altering Soil Reaction: Adjusting soil pH to make it less acidic or less alkaline.
2. Enhancing Nutrient Availability: Converting nutrients in the soil from unavailable to readily available forms for plant uptake.
3. Improving Physical Properties: Enhancing the physical structure of the soil, such as porosity and water-holding capacity.
4. Mitigating Harmful Substances: Neutralizing the impact of toxic substances present in the soil.

Many soil amendments also supply essential nutrients to plants. For instance, agricultural liming materials provide calcium and, in some cases, magnesium, which are vital for plant growth.

Types of Soil Amendments

Soil amendments can be broadly categorized into three types based on their primary function:

1. Materials for Correcting Soil Acidity: These are used to neutralize acidic soils, typically found in high-rainfall regions where bases are leached out.
2. Materials for Correcting Soil Alkalinity: These amendments are applied to saline and alkaline soils, common in arid and semi-arid areas.
3. Soil Aggregating Agents or Conditioners: These substances improve soil structure by stabilizing soil aggregates and promoting the formation of a granular soil structure.

Soil Reaction and pH

The reaction of a soil, whether acidic or alkaline, is measured on the pH scale and plays a crucial role in determining soil fertility. High-rainfall regions often experience the formation of acidic soils due to base leaching, whereas arid and semi-arid regions tend to have saline and alkaline soils due to limited leaching.

Impact of Soil Acidity on Fertility

Soil acidity significantly affects nutrient availability and often leads to mineral stress, which can impair plant growth. The infertility of acidic soils is a complex issue arising from multiple interrelated factors. These include:

1. Nutrient Deficiencies: Acidic conditions can reduce the availability of essential nutrients such as phosphorus, calcium, and magnesium.
2. Toxicities: High concentrations of certain elements, such as aluminium and manganese, become toxic to plants in low pH soils.
3. Impaired Root Development: Acidic conditions can hinder root growth, limiting the plant's ability to absorb water and nutrients.

Addressing soil acidity and alkalinity through appropriate amendments is essential for improving soil fertility and ensuring sustainable crop production.

Benefits of Liming

Proper application of lime to acidic soils offers several agronomic and environmental benefits, as outlined below:

1. Improved Nutrient Solubility and Availability: Raising soil pH through liming enhances the solubility and availability of most nutrients, with the exception of molybdenum (Mo), which becomes more available. At higher pH levels (above 6.0), however, certain micronutrients may become less soluble, potentially leading to deficiencies.
2. Enhanced Root Development: Liming reduces the toxicity of aluminium (Al) and manganese (Mn), which promotes better root growth. This, in turn, facilitates greater nutrient uptake and improves the plant's tolerance to drought.
3. Source of Calcium and Magnesium: Lime acts as an economical source of calcium, and if dolomitic limestone is used, it also supplies magnesium—both essential nutrients for plant growth.
4. Improved Phosphorus Use Efficiency: Liming reduces phosphorus fixation by active aluminium and iron oxides in the soil, making native and applied phosphorus more available to plants.
5. Increased Cation Exchange Capacity (CEC): An increase in soil pH enhances the soil's CEC, reducing the leaching of essential base cations like potassium, thereby improving nutrient retention.
6. Enhanced Nodulation in Legumes: Liming improves nitrogen fixation by promoting nodulation in legume roots. This is facilitated by the activity of Rhizobium bacteria, which thrive in optimal pH conditions.
7. Molybdenum Availability: As soil pH increases with liming, the availability of molybdenum—an essential nutrient for *Rhizobium* in nitrogen fixation—also improves.

8. Herbicide Degradation: Optimal pH conditions help in the breakdown of certain herbicides, preventing potential damage to subsequent crops in rotational farming systems.
9. Enhanced Microbial Activity: Liming fosters a favorable environment for soil microbes, increasing organic matter decomposition and nutrient cycling. This, in turn, supports healthier root systems.
10. Improved Earthworm Activity: Deep-burrowing and soil-ingesting earthworm species perform better in soils without acid stress, contributing to enhanced soil structure and fertility.

Common Liming Materials

The most commonly used agricultural liming materials include:

- Ground Limestone (Calcium Carbonate, $CaCO_3$)
- Burnt Lime (Calcium Oxide, CaO)
- Hydrated Lime (Calcium Hydroxide, $Ca(OH)_2$)

These materials vary in their chemical composition and effectiveness, providing flexibility in addressing specific soil acidity challenges.

Types and Characteristics of Liming Materials

Among the three forms of liming materials, ground limestone is the most widely used across various countries for soil amendment purposes.

Ground Limestone

Limestone refers to a type of rock primarily composed of calcium carbonate ($CaCO_3$) or a mixture of calcium and magnesium carbonate ($CaMg(CO_3)_2$), known as dolomitic limestone. Extensive limestone deposits are present in nearly all states of India. Currently, Odisha is the largest producer of limestone, followed by Bihar, where most of it is utilized for cement production. The limestone from Gujarat is regarded as one of the finest, with a calcium carbonate content of approximately 92%. When applied to acidic soils, ground limestone initiates a series of chemical reactions. A portion of calcium carbonate dissolves in the soil and reacts with carbonic acid to form calcium bicarbonate, as shown in the following reaction:

Reaction 1

$$CaCO_3 + H_2CO_3 \rightarrow Ca(HCO_3)_2$$

(Limestone) + (Carbonic acid) → (Calcium bicarbonate in solution)

The calcium bicarbonate solution then interacts with soil colloids, liberating carbon dioxide:

Reaction 2

$$\overset{\text{H}}{\underset{\text{H}}{\boxed{\text{Micelle}}}} + Ca(HCO_3)_2 \rightarrow Ca\text{-}\boxed{\text{Micelle}} + H_2O + CO_2$$

(Soil colloid) + (Calcium bicarbonate) → (Calcium colloid complex) + Carbon dioxide + Water

In addition to its solution-based reactions, a portion of the limestone remains in solid form and comes into direct contact with soil colloids. This solid-state interaction results in the absorption of calcium by soil colloids, further contributing to soil pH modification:

Reaction 3

$$\overset{\text{H}}{\underset{\text{H}}{\boxed{\text{Micelle}}}}\ CaCO_3\ (\text{solid}) \rightarrow Ca + \boxed{\text{Micelle}} + H_2O + CO_2$$

Through these reactions, carbon dioxide is consistently released, and the absorbed calcium and magnesium increase the base saturation of the soil colloidal complex. This process effectively raises the soil pH, mitigating acidity.

The quality and effectiveness of ground limestone as a soil amendment depend on two key factors:

- Chemical Composition: The proportion of calcium and magnesium content.
- Fineness: The degree of particle size reduction, as finer particles dissolve more readily in soil.

By-Product Lime: In addition to natural sources, lime is also produced as a by-product in industries such as paper manufacturing, sugar refining, and ammonium sulphate production. These by-product limes consist mainly of chemically precipitated calcium carbonate in a fine form, making them highly effective as soil amendments. For instance, the Sindri Fertilizer Factory generates lime as a by-product during the production of ammonium sulphate through the gypsum process. By-product lime provides an economical alternative to natural limestone while maintaining similar efficacy in neutralizing soil acidity and enhancing soil health.

Burnt lime

Burnt lime, also referred to as quick lime, lime oxide, caustic lime, or stone lime, is produced by heating limestone ($CaCO_3$) to temperatures exceeding 800°C. During this process, calcium carbonate decomposes into calcium oxide (CaO) and carbon dioxide (CO_2), as shown in the reaction:

Reaction 1

$CaCO_3 \rightarrow CaO + CO_2$

(Limestone) → (Calcium oxide) + (Carbon dioxide)

For every 100 kg of calcium carbonate heated, 56 kg of calcium oxide is generated. Similarly, burning 84.3 kg of magnesium carbonate produces 40.3 kg of magnesium oxide. The purity of burnt lime used in agriculture varies, typically ranging between 85% and 98%. When added to acidic soil, burnt lime reacts with water to form calcium hydroxide (slaked lime), which further reacts with carbonic acid to produce calcium bicarbonate in solution:

Reaction 2

$CaO + H_2O \rightarrow Ca(OH)$

(Burnt lime) + (Water) → (Hydroxide of lime)

Reaction 3

$Ca(OH)_2 + 2H_2CO_3 \rightarrow Ca(HCO_3)_2 + 2H_2O$

(Hydroxide of lime) + (Carbonic acid) → (Calcium bicarbonate) + (Water)

The calcium bicarbonate interacts with soil colloids, leading to the adsorption of calcium ions, which increases the base saturation and soil pH:

Reaction 4

$\text{Micelle} + Ca(HCO_3)_2 \rightarrow Ca + 2CO_2 + \text{Micelle}\ \ 2H_2O$

(Soil colloid) + (Calcium bicarbonate) → (Calcium colloid complex) + Carbon dioxide + Water

Slaked Lime

Slaked lime, also called hydrated lime or calcium hydroxide, is produced by adding water to burnt lime:

Reaction 5

$CaO + H_2O \rightarrow Ca(OH)_2$

(Burnt lime) + (Water) → (Slaked lime)

For every 56 kg of calcium oxide, approximately 18 liters of water are required to produce 74 kg of slaked lime. The process generates significant heat, and the final product is a whitish gray powder marketed as agricultural lime. When slaked lime is applied to acidic soil, it reacts with carbonic acid to form calcium bicarbonate, as follows:

Reaction 6

$Ca(OH)_2 + 2H_2CO_3 \rightarrow Ca(HCO_3)_2 + 2H_2O$

(Slaked lime) + (Carbonic acid) → (Calcium bicarbonate) + (Water)

The calcium bicarbonate then interacts with soil colloids, enhancing base saturation and soil pH while releasing carbon dioxide:

Reaction 7

$$\begin{matrix} H \\ \text{Micelle} \\ H \end{matrix} + Ca(HCO_3)_2 \rightarrow Ca\ \text{Micelle} + 2CO_2 + 2H_2O$$

Comparative Efficiency of Liming Materials

The effectiveness of liming materials in correcting soil acidity depends on the following factors:

1. Calcium Carbonate Equivalent (Neutralizing Value): The neutralizing value (N.V.) represents the material's capacity to neutralize acidity, expressed as a percentage of pure calcium carbonate ($CaCO_3$), which has an N.V. of 100%. For burnt lime (CaO), the calcium carbonate equivalent is calculated as:

$$CaCO_3 Equivalent = \frac{Molecular\ weight\ of\ CaCO_3}{Molecular\ weight\ of\ caC} x100$$

$$= \frac{100}{56} \times 100$$

$$= 178.6$$

2. Purity: The effectiveness of a liming material is significantly influenced by its chemical purity. For instance, if ground limestone contains 95% calcium carbonate ($CaCO_2$), its neutralizing value will also be 95%. Materials with higher purity are more efficient in neutralizing soil acidity.
3. Degree of Fineness: The fineness of a liming material is equally critical because it determines how quickly the material reacts with acidic soil. Finer particles dissolve faster and are thus more effective. The fineness of a material is measured by its ability to pass through sieves of specified mesh sizes. A 60-mesh sieve, with 60 equal-sized holes per linear inch, is commonly used as a standard. In terms of efficiency rating, materials passing through a 60-mesh sieve are considered to have 100% effectiveness. Materials that pass through a 20-mesh sieve but not a 60-mesh sieve are assigned a 60% efficiency rating, while those passing through an 8-mesh sieve but not a 20-mesh sieve are rated at 20% efficiency.

Efficiency Ratings (Ohio Scale)

- Material passing through a 60-mesh sieve: 100%
- Material passing through a 20-mesh but not a 60-mesh sieve: 60%
- Material passing through an 8-mesh but not a 20-mesh sieve: 20%

Factors Influencing Lime Requirement

The amount of lime needed to correct soil acidity depends on several factors:

1. Soil pH (Intensity of Acidity): Strongly acidic soils require higher amounts of lime compared to slightly acidic soils to neutralize acidity effectively.
2. Cation-Exchange Capacity (CEC): Lime requirements are directly influenced by the soil's cation-exchange capacity, which is determined by the quantity and type of clay minerals and organic matter. Soils with a higher CEC require more lime to achieve a similar increase in pH compared to soils with lower CEC.
3. Soil Texture: The texture of the soil also affects lime requirements. Sandy soils typically need less lime than finer-textured soils such as silt loam or clay loam, which have higher buffering capacities.
4. Lime Requirement of Crops in Rotation: Pulses and legumes significantly benefit from lime application, as they do not perform well in strongly acidic soils. These crops are also high consumers of lime. However, for crops like potatoes, lime application should be carefully managed. The organism responsible for potato scab thrives in moderately acidic to alkaline soils, while its activity is suppressed in sufficiently acidic conditions. Therefore, a lower lime dose is recommended for potato cultivation to prevent the development of potato scab.
5. Purity of Liming Material: The purity of the liming material directly affects its efficiency. Higher purity materials require lower quantities to achieve the desired neutralization of soil acidity.
6. Chemical Composition of Liming Material: The chemical composition of the liming material influences its neutralizing value. Calcium carbonate is used as the standard, with a neutralizing value of 100. Consequently, smaller quantities of higher-reactivity materials, such as burnt lime, are needed compared to ground limestone.
7. Fineness or Particle Size: The effectiveness of a liming material is greatly enhanced by its fineness. Finer materials react more quickly with the soil, leading to improved efficacy in neutralizing acidity.

   Lime Requirements for Indian Soils: Studies on lime requirements for Indian soils are limited. However, for the acidic soils of hilly regions in Punjab and Himachal Pradesh (Table 1), specific lime application schedules have been proposed to address soil acidity and improve crop productivity.

Table 1: Lime requirements of acid soils of Punjab and Himachal Pradesh

	Limestone required (kg/ha) for different textual classes to raise their pH to 6.5		
Soil pH	**Sandy loam**	**Loam**	**Clay loam**
5.0	1260	1889	2940
5.2	1092	1637	2548
5.4	924	1385	2156
5.6	756	1133	1764
5.8	588	881	1372
6.0	420	629	980
6.2	252	377	588

Methods of Lime Application

The following principles should be followed when applying lime:

1. Lime should be applied either before ploughing or on ploughed land, followed by thorough mixing into the soil using a disc harrow.
2. When large amounts of lime are required, it is better to divide the total amount into two or three smaller applications rather than applying it all at once. For instance:
    - On strongly acidic soils requiring 3–6 tonnes of lime per hectare, half the quantity should be applied before ploughing, and the remaining half incorporated after ploughing.
    - When the lime requirement is up to 2 tonnes per hectare, the entire amount can be applied at once and mixed into the soil 8–10 days before sowing.
3. Lime should not be applied immediately before ploughing with a furrow-turning plough if only a single application is planned.

Industrial Wastes as Amendments for Acid Soils

Several industrial by-products have proven effective as soil amendments for acidic soils. These include: Basic slag and blast furnace slag from steel industries, lime sludge from paper mills, precipitated calcium carbonate from fertilizer factories, cement kiln wastes, press mud and cake from sugar mills. These industrial by-products provide cost-effective alternatives to conventional liming materials and contribute to sustainable soil management practices.

Saline and Alkaline Soils

In the arid and semi-arid regions of India, some soils exhibit high concentrations of soluble salts, exchangeable sodium, or both. These soils pose challenges to crop production and are categorized as problematic soils requiring specific management practices. Excessive levels of soluble salts can adversely affect plant growth, rendering the soil unsuitable for crop cultivation when salt accumulation becomes severe. Soils with high salt content typically have a pH greater than 7.0 and are classified into the following categories based on their pH range:

- Mildly alkaline: pH 7.4 to 7.8
- Moderately alkaline: pH 7.9 to 8.4
- Strongly alkaline: pH 8.5 to 9.0
- Very strongly alkaline: pH 9.1 and higher

Sources of Soluble Salts

The salts that accumulate in saline and alkaline soils primarily include:

- **Cations**: Calcium (Ca^{2+}), magnesium (Mg^{2+}), and sodium (Na^{+}).
- **Anions**: Chloride (Cl^{-}) and sulfate (SO_4^{2-}) in large amounts, along with bicarbonate (HCO_3^{-}), carbonate (CO_3^{2-}), and nitrate (NO_3^{-}) in smaller amounts.

The predominant salts found in these soils are sodium chloride, sodium sulphate, calcium chloride, calcium sulphate, magnesium sulphate, and magnesium chloride. Minor amounts of sodium nitrate, sodium bicarbonate, and calcium bicarbonate may also be present. The specific composition of these salts determines whether the soil is classified as saline or alkaline.

Formation of Saline and Alkaline Soils

In low-rainfall regions, the salts produced by the weathering of soil minerals are not sufficiently leached. During periods of higher-than-average precipitation, soluble salts may be leached into the subsoil, where poor drainage leads to their accumulation. As water in the soil evaporates, it leaves behind concentrated salt deposits. This cycle of leaching and deposition gradually results in toxic levels of salts in the root zone, severely affecting plant growth. Additionally, sodium in the soil can replace calcium in the clay complexes, further degrading soil structure and fertility. Over time, calcium is removed, and sodium dominates, making the soil highly alkaline. The chemical reactions contributing to this process include:

Calcium clay + $2NaCl \rightarrow$ Sodium clay + $CaCl_2$ (leachable calcium chloride)

Calcium clay + $Na_2SO_4 \rightarrow$ Sodium clay + $CaSO_4$ (leachable calcium sulfate)

Calcium clay + $Na_2CO_3 \rightarrow$ Sodium clay + $CaCO_3$ (calcium carbonate deposition)

Classification of Saline and Alkaline Soils

1. Saline Soils

- Contain high levels of soluble salts that reduce crop productivity but lack excess exchangeable sodium.

- pH is typically less than 8.5.
- Often identified by white salt crusts on the soil surface, primarily composed of calcium and magnesium chlorides, sulfates, and occasionally nitrates. These soils are also referred to as "white alkali soils."

2. Saline-Alkali Soils

- Contain both significant soluble salts and high levels of exchangeable sodium, which impair soil properties and hinder plant growth.
- pH exceeds 8.5.

3. Non-Saline-Alkali Soils

- Characterized by excessive exchangeable sodium but low soluble salt concentrations.
- pH usually ranges from 8.5 to 10.

Amendments for Reclaiming Saline and Alkaline Soils

Reclaiming saline and alkaline soils involves incorporating chemical amendments into the soil to enhance its properties and improve productivity. These amendments primarily serve two purposes: (1) providing soluble calcium to replace exchangeable sodium and (2) neutralizing alkaline salts through acid formation. The following amendments are commonly used, depending on the specific soil conditions:

1. Gypsum ($CaSO_4.2H_2O$): Gypsum, a naturally occurring calcium sulphate, is abundant in regions such as Rajasthan. It reacts with exchangeable sodium in the soil, forming sodium sulphate, which can be leached out to reduce soil pH. The chemical reaction is: $CaSO_4 + 2NaX \rightarrow CaX + Na_2SO_4$

   Gypsum + Soil complex saturated with sodium → Soil complex saturated with calcium + Sodium sulphate

   Here, "X" represents the soil complex. Gypsum is typically applied at rates of 1–2 tonnes per hectare but may be used up to 20 tonnes per hectare depending on the soil pH and the concentration of exchangeable sodium.
2. Sulphur (S): Sulphur is an effective amendment for replacing exchangeable sodium with calcium. Under field conditions, one atom of sulphur can replace approximately three sodium ions in the soil complex. The process involves oxidation of sulphur to sulphuric acid, which reacts with soil calcium carbonate to form calcium sulphate. The reaction sequence is:

   $2S + 3O_2 \rightarrow 2\ SO_3$ (Microbiological oxidation)

   Sulphur trioxide

   $2SO_3 + H_2O \rightarrow 2\ H_2SO_4$

   Sulphuric oxide

$H_2SO_4 + CaCO_3 \rightarrow CaSO_4 + CO_2 + H_2O$

$2NaX + CaSO_4 \rightarrow CaX_2 + Na_2SO_4$

Sulphate should be thoroughly mixed into the soil and allowed to react several weeks before planting. The required amount depends on soil properties and climatic conditions.

3. Iron sulphate ($FeSO_4$): Iron sulphate can also be used to reclaim alkali soils. It forms sulfuric acid upon reaction with water, which subsequently reacts with soil calcium carbonate to produce calcium sulphate. The calcium replaces exchangeable sodium in the soil complex. The reaction steps are as follows:

   $FeSO_4 + H_2O \rightarrow H_2SO_4 + FeO$

   $H_2SO_4 + CaCO_3 \rightarrow CaSO_4 + CO_2 + H_2O$

4. Limestone ($CaCO_3$): Ground limestone is effective for soils with pH levels ranging from 7.0 to 7.5. The calcium component of limestone reacts with the soil complex, replacing sodium and forming sodium carbonate, which is subsequently leached out. The reactions are:

   $CaCO_3 + 2NaX \rightarrow CaX + Na_2CO_3$

   However, limestone is ineffective in soils with pH levels exceeding 7.5 due to its reduced solubility at higher pH levels.

5. **Pyrites (FeS_2)**: Pyrites are sometimes used as a source of sulphur for alkali soil reclamation. Upon oxidation, pyrites release sulphuric acid, which can facilitate the replacement of exchangeable sodium with calcium in the soil complex.

Application Guidelines

The choice of amendment and application rates depend on soil pH, exchangeable sodium levels, and climatic conditions. Proper mixing of amendments into the soil is crucial to ensure uniform reaction. In addition, amendments should be applied well in advance of planting to allow sufficient time for chemical reactions and leaching processes.

Practicals

1

Determination of Available Potassium in Soils

Available potassium

The term "available potassium" refers to the total of exchangeable and water-soluble potassium. These two forms of potassium make up only about 1-2% of the total potassium in soil. They exist in equilibrium with non-exchangeable potassium, which is potassium that is fixed by minerals such as vermiculite, illite, and other 2:1 type clay minerals found in the soil. When crops absorb exchangeable potassium or when potassium is leached away, some of the fixed potassium is released to maintain this equilibrium, thus becoming available for plant uptake. Additionally, potassium present in the form of minerals like feldspars and micas is also significantly important for overall potassium availability in the soil. To determine the release of potassium from non-exchangeable forms, an extraction method using boiling 1N HNO_3 (Wood and De Turk in 1941) has been proposed, which can serve as an index of potassium availability in soils. However, the most widely accepted measure of potassium availability is the sum of exchangeable and water-soluble potassium. This extraction can be performed using a neutral normal ammonium acetate solution, a method introduced by Merwin and Peech in 1950. Further details of this method are provided below:

Apparatus: Conical flasks (150 ml)- 2 nos, Balance, Pipette (25 ml), Funnel with funnel stand, Electric shaker, Volumetric flasks (25 ml)- 8nos, Wash bottle, Flame photometer

Reagents: 1N solution of ammonium acetate (pH 7.0): Dissolve 77.09 g of ammonium acetate (CH_3COONH_4) in distilled water and dilute to 1 liter.

Standard solution of K (100 ppm): Dissolve 0.1907 g of KCl (AR Grade) in distilled water and dilute to 1 liter.

Standard solution of K (10 ppm): Dilute the 100 ppm K solution by 10 times.

Filter paper: Whatman No.1

Principle: Soil is shaken with neutral normal ammonium acetate and the suspension is filtered. The extract, thus obtained is tested for its content of potassium with the help of a flame-photometer. Ammonium ions of the extractant replace the potassium ions which are thus brought into the solution. One of the advantages of this extract is that the ammonium (NH_4^+) ions help in reducing the contamination due to non-exchangeable potassium as the former has almost the same ionic radi as that of K^+.

Reaction

$$\boxed{\text{Soil Colloid}}\begin{matrix} K+ \\ \\ K+ \end{matrix} + 2CH_3COONH_4 \longrightarrow \boxed{\text{Soil Colloid}}\begin{matrix} NH_4 \\ \\ NH_4^+ \end{matrix} + 2CH_3COOK$$

Procedure

Weigh 5 g of soil in a 150-ml conical flask, add 25 ml of neutral ammonium acetate solution and shake for five minutes. Filter the suspension through Whatman No.1 filter paper. Pipette 5 ml of this extract into 25 ml volumetric flask and make its volume with distilled water. Feed this solution to the atomizer of the flame photometer in which 100 of the Galvanometer has been set by feeding 10 ppm solution of K to the atomizer and record the reading indicated by the galvanometer needle. Then with the help of the standard curve (which can be prepared as given) calculate the amount of available potassium in the soil under test.

Preparation of standard curve of K

Pipette 0, 0.5,1.0, 1.5, 20 and 2,5 ml of 100 ppm K solution into 25 ml volumetric flask, add 5 ml neutral normal ammonium acetate solution and make the volume with distilled water. Set the 100 of galvanometer (zero of which has been set with blank) of the flame photometer with 10 ppm solution of K. Then feed the solutions of different concentration to the atomizer and note reading of the galvanometer on simple graph paper obtained for every concentration. Plot a curve between the concentration of K and reading.

Precautions

1. The air pressure should neither be below 0.4 kg/cm nor it should exceed 0.75 kg/cm.
2. The flame should be soot free and blue.
3. While starting the instrument turn on the air supply before turning on the fuel supply.
4. When extinguishing the flame, always turn off the fuel before turning off the air.
5. 10-15 minutes must be given for the instrument to warm up.
6. When all the test samples have been analyzed, the atomizer must be fed with distilled water continuously for about 2 minutes.
7. Shake the suspension exactly for 5 minutes.
8. Before analyzing the extract, ensure that it is quite clear.

Calculations

Weight of the soil taken =5g

Vol. of the extractant added =25 ml

1st Dilution = 5 times

2nd dilution = 5 times

Total dilution = 25 times

Reading indicated by the galvanometer = X

ppm of K as read from the standard Curve against X= Y

ppm of K in the given soil = Y x 25 = Z

Kg of K/ha = Zx2.24

Interpretation

Amounts of available K (kg/ha)	Rating
<136	Low
136-333	Medium
>333	High

References

Black, C.A. (1965). Methods of soil analysis. Part 2. Chemical and Microbiology Properties. Pp: 489-856.

Merwin, H.D. and Peech, M. (1950). Exchangeability of soils potassium in the sand, silt and clay fractions as influenced by the nature of the complementary exchangeable cations. Soil Sci. Soc. Amer. Proc. 15: 125-28.

Jackson, M.L. (1967). Soil Chemical analysis. Pp 452-85 039 0.

2

Determination of Available Micronutrients in Soil

Micronutrient cations (Zinc, copper, Iron and Manganese)

The imbalanced use of high analytical inorganic fertilizers, along with minimal or no use of organic manures and significant removal of micronutrients, has led to deficiencies of these nutrients in the soil. Micronutrient deficiencies are widespread in Indian soils, with an average of 43.0%, 12.1%, 5.4%, and 5.6% of soils being deficient in zinc (Zn), iron (Fe), copper (Cu), and manganese (Mn), respectively. Micronutrients are as critical to plant nutrition as primary and secondary nutrients. The deficiency of any one micronutrient can limit plant growth, even when other essential nutrients are present in adequate amounts. The relative proportions of micronutrients in the soil, particularly metals, influence their availability and may be more significant than the absolute amounts of each nutrient. The total micronutrient content in soils varies widely, ranging from 2 to 1600 ppm for Zn, 2000 - 100000 ppm for Fe, 2 to 960 ppm for Cu, and 37 to 4600 ppm for Mn. Soil analysis using various extractants can assess soils; ability to supply micronutrients, as detailed in Table 1. Research indicates that chelating agents are effective for evaluating the readily available micronutrient cations in the soil. These agents form soluble complexes with free metal ions in solution, thereby reducing the activity of free metal ions. As a result, metal ions desorb from soil surfaces or dissolve from labile solid phases to replenish the free metal ions in the soil solution.

Table 1: Extractants used for assessing soil micronutrient status.

Element	**Extractant**	**Element**	**Extractant**
Zn	EDTA + Ammonium carbonate	Cu	EDTA
	EDTA + Ammonium acetate		EDTA + Ammonium acetate
	DTPA + $CaCl_2$		Ammonium bicarbonate + DTPA
	HNO_3		HCl
	HCl		HNO_3
	Dithizone + Ammonium acetate		DTPA + $CaCl_2$

Element	Extractant	Element	Extractant
Fe	EDTA		Hydroquinine + Ammonium acetate
	EDTA + Ammonium acetate		DTPA + $CaCl_2$
	HCl		HCl + H_2SO_4
	HNO_3		HCl
	Ammonium bicarbonate + DTPA		HNO_3
	DTPA + $CaCl_2$		EDTA + Ammonium acetate

DTPA-$CaCl_2$ -TEA Extraction Method for Available Metallic Ions (Zn, Fe, Cu, Mn)

The DTPA-$CaCl_2$ -TEA extraction method, developed by Lindsay and Norvell in 1978, is widely recognized and commonly used for determining the availability of micronutrients in soil samples. This method utilizes DTPA (diethylene triamine penta acetic acid) as an extractant, which is well-regarded for simultaneously extracting essential micronutrient cations, such as zinc (Zn), iron (Fe), copper (Cu), and manganese (Mn), particularly in neutral and alkaline soils. The concentrations of these cations in the extract are measured using an Atomic Absorption Spectrophotometer (AAS) equipped with specific hollow cathode lamps. Further details of this method are provided below:

Principle: The four micronutrient cations—zinc (Zn), copper (Cu), iron (Fe), and manganese (Mn)—are extracted from soil by shaking it with a DTPA extracting solution. This solution consists of 0.005 M DTPA, 0.1 M triethanolamine (TEA), and 0.01 M $CaCl_2 \cdot 2H_2O$, buffered at pH 7.3. DTPA acts as a chelating agent, binding with free metal ions in the solution to form soluble complexes. The stability constants for the simultaneous complexing of Zn, Fe, Cu, and Mn indicate that DTPA is a highly suitable extractant. DTPA has the capacity to complex each of the micronutrient cations at levels up to 10 times their atomic weights, with capacities ranging from 550 to 650 ppm, depending on the specific cation. During the extraction process, TEA becomes protonated as $HTEA^+$, which facilitates the release of micronutrient cations from the solid phase into the solution, where they are then chelated by DTPA. Buffering the extracting solution in a slightly alkaline pH range and including soluble Ca^{2+} through $CaCl_2 \cdot 2H_2O$ helps prevent the dissolution of $CaCO_3$. This exclusion is important as it ensures that the extraction only measures the available micronutrients, excluding those occluded forms that are not accessible for absorption by plant roots. Finally, the concentrations of the extracted micronutrient cations in the solution are measured using an atomic absorption spectrophotometer.

Apparatus/Equipment

Analytical balance, 100 ml capacity narrow mouth polyethylene bottles, pipettes, reciprocating electric shaker, Whatman No. 1 or 42 filter paper, polyethylene/ glass funnels, polyethylene vials of 50 ml capacity, Atomic Absorption Spectrophotometer (AAS) and hollow cathode lamps of Zn, Cu, Fe and Mn.

Reagents

1. Extracting solution: The solution is prepared to contain 0.005 M DTPA, 0.01 M $CaCl_2.2H_2O$ and 0.1 M TEA (Triethanolamine), and its pH is adjusted to 7.3. Dissolve 1.967 g DTPA and 13.3 ml TEA in deionzed water. Add 1.47 g $CaCl_2.2H_2O$ in about 500 ml deionized water to a one-litre volumetric flask, add the DTPA—TEA mixture to it, and make the final volume about 900 ml. Adjust pH to 7.3 using 6N HCl, make the final volume to one litre, and mix thoroughly. This solution is stable for several months.
2. Stock standard solutions (1000 mg metal^{-1}): The stock standard solutions containing 1000 mg of the given micronutrient per litre can be prepared by dissolving the required quantity of their respective salts (as shown in Table 2) and making the final volume to one litre with deionized water.

Element	Concentration of stock (µg/ml)	Salt to be used (chemical formula)	Quantity of salt to be dissolved in one litre of solution (g)
Zn	1000	Zinc sulphate ($ZnSO_4.7H_2O$)	4.398
Cu	1000	Cupric sulphate ($CuSO_4.5H_2O$)	3.928
Fe	1000	Ferrous sulphate ($FeSO_4.7H_2O$)	4.964
Mn	1000	Manganese sulphate ($MnSO_4.H_2O$)	3.076

Working standard solutions: From 1000 mg l^{-1} stock solution of Zn, Cu, Fe and Mn; prepare 50 mg l^{-1} working solution. For this purpose, pipette out 5 ml of 1000 mg l^{-1} solution in a 100 ml volumetric flask and make up the volume with the deionized water. From this solution, prepare the required standard solutions of each element in a 50 ml volumetric flask by pipetting out the required volume of the working standard and making the final volume with 0.005M DTPA. The concentration ranges usually required are 0, 0.2, 0.4, 0.8 and 1.2 mg l^{-1} for Zn; 0, 0.2, 0.4, 0.8 and 1.0 mg l^{-1} for Cu; 0, 2, 4, 6, 8 and 10 mg l^{-1} for Fe and 0, 1, 2, 3, and 4 mg l^{-1} for Mn.

Procedure

A. Extraction of soil samples

i. Weigh 10 g of thoroughly processed air-dried soil sample.
ii. Transfer it to a 100 ml narrow-mouth polyethylene bottle.

iii. Add 20 ml DTPA-extracting solution.
iv. Stopper the bottle and shake on an electric shaker for exactly two hours at 25°C
v. Filter the contents of the flask using Whatman No. 1 or 42 filter paper. Ensure that filtrate is free of colloidal matter.
vi. Also, run a blank with only DTPA solution and no soil.

B. Analysis of extracts

The soil extracts prepared through the prescribed procedure should be used to determine the content of micronutrient cations (Zn, Fe, Cu and Mn) using an atomic absorption spectrophotometer. The procedure is described here under:

i. Set zero of the instrument with blank.
ii. Feed standards belonging to the elements are to be determined by the atomic absorption spectrophotometer to standardize the instruments to read absorbance and/or concentration in the samples having the given element within the standardized range.
iii. Then feed the DTPA extract (s) and record the absorbance/ concentration of the elements in question.
iv. Repeat the above steps for every element.
v. In case the instrument shows a sign of 'over' for some element in a particular sample, indicating thereby that the sample has a concentration out of the range for which the instrument has been standardized. In such a situation, the extract has to be diluted further and fed again to the instrument for recording the absorbance.

Calculations

Most modern AASs are calibrated to display the concentration of a micronutrient in ppm directly in the soil extract. In such cases, the concentration of the given micronutrient in the soil sample is calculated by multiplying the displayed reading by the dilution factor.

Available (DTPA-extractable) Zn/Fe/Cu/Mn is soil (ppm) = $A \times B / C$

Where, A= sample reading obtained from AAS;

B = Weight of DTPA extractant in ml;

C = Weight of soil sample in g.

Interpretation

Soils with DTPA extractable Zn, Cu, Fe, and Mn less than 0.6, 0.2, 4.5, and 3.5 mg kg^{-1}, respectively, are said to be deficient in the respective micronutrient, whereas those with more than these values are considered to have an adequate supply of these micronutrients.

Precautions

1. Ensure that the deionized water is used during the analysis.
2. The apparatus (glass/polyethylene) for analysis must be thoroughly washed with acidified water and then with deionized water.
3. Shaking must be carried out at 25°C.
4. Analytical grade reagents should be used to avoid contamination due to impurities.
5. Before feeding the extracts, it must be ensured that they are not turbid; otherwise, they may block the capillary of the AAS.

3

Determination of Available Sulphur in soils

Sulphur is present in soils both in organic and inorganic forms. The amount depends upon parent material, climate, soil organic matter status, soil texture, etc. In humid region surface soils, most of the sulphur is in organic form, but in arid region soils, sulphates of Ca, Mg, Na, and K are present in large quantities in the soil profile. Plants absorb sulphur in the form of sulphate ions (SO_4^{2-}). A large number of extractants like water, monocalcium phosphate [$Ca(H_2PO_4)_2$], Calcium Chloride ($CaCl_2$), potassium dihydrogen orthophosphate (KH_2PO_4), Morgan's reagent (CH_3COO-Na-CH_3-COOH), (ammonium acetate) $CH_3COONH_4^+$ (acetic acid), CH_3COOH, and NaCl have been used to extract available S. The 0.15% $CaCl_2$ extraction method suggested by Williams and Steinbergs (1959) is the most widely used among different extraction procedures. The details of this method are given below:

Principle

Soil is shaken with 0.15% $CaCl_2$ solution. Chloride ions displace adsorbed sulphate during extraction. Calcium ions suppress the extraction of soil organic matter and hence eliminate the contamination caused by extractable organic S. The filtrate is analyzed for S by the turbidimetric method of Chesnin and Yien (1950) in which the turbidity produced due to the precipitation of sulphate as barium sulphate is measured on a Colorimeter at a wavelength of 420 nm using a blue filter. Gum acacia solution is added to stabilize the turbidity so the barium sulphate precipitates do not settle.

Reactions

$CaCl_2 \longrightarrow Ca^{++} + 2Cl^-$

Micelle] $SO_4^{-2} + 2Cl^- \longrightarrow$ Micelle] $Cl^- + SO_4^{2-}$

$BaCl_2 \longrightarrow Ba^{2+} + 2Cl^-$

$SO_4^{2-} + Ba^{2+} \longrightarrow BaSO_4$ (Precipitate)

Reagents

1. Extracting solution: Dissolve 1.986 g of calcium chloride dihydrate ($CaCl_2.2H_2O$) in distilled water and dilute to one litre.
2. Barium chloride ($BaCl_2$); 30—to 60-mesh crystals: Grind barium chloride crystals until they pass through a 30 -mesh sieve and are retained on a 60-mesh sieve.
3. Gum acacia solution (0.25%): Dissolve 0.25 g gum acacia in distilled water and dilute to 100 ml.
4. Standard sulphur solution (100 ppm S): Dissolve 0.5434 g of the reagent grade potassium sulphate (K_2SO_4) in distilled water and dilute to one litre.

Apparatus

150-ml Erlenmeyer flask, funnel, pipettes, 25-ml volumetric flasks, electric Shaker, colorimeter and Whatman No.42 filter paper.

Procedure

a) Preparation of standard curve: Pipette out 0.25, 0.5, 1.0. 2.5 and 5.0 ml of 100 ppm S solution in different 25-ml volumetric flasks. Add 10 ml of 0.15% $CaCl_2$ solution to every flask and 1.0 g of 30-60 mesh $BaCl_2$ crystals. Swirl for one minute to dissolve the crystals, and add 1 ml of 0.25% gum acacia solution. Make up the volume in every flask with distilled water and shake well. Within 5 to 30 minutes after the development of turbidity, read the standards on a colorimeter at 420 nm using a blue filter. Plot a standard curve showing the relationship between the concentration of S and turbidity/absorbance readings.

b) Analysis of test sample: Weight 10 g of air-dried soil and transfer it to a 150-ml conical flask. Add 50 ml of 0.15 % $CaCl_2$ solution and shake for 30 minutes on an electric shaker. Filter the suspension through the Whatman No.42 filter paper. Pipette 20 ml of the filtrate in a 25-ml volumetric flask and proceed as in the case of the standard curve. Run a blank with all the chemicals except the soil. Find out the sulphate concentration from the standard curve.

Calculations

Weight of soil taken = 10g

The volume of extractant added = 50

First dilution = 5 times

The volume of aliquot taken = 20 ml

Final volume = 25 ml

Second dilution = 1.25 times

Total dilution =5 x 1.25 = 6.25 times

Absorbance reading = A

ppm of S from the standard curve against A value = Y

ppm of S in soil = Y x6.25 ppm

Kg of S/ha = ppm x2.24

Inference

Soils with S available at less than 10 ppm are rated deficient in sulphur.

References

Chesnin, L. and C.H. Yien. (1950). Turbidimetric determination of available sulphates. Soil Sci. Soc. Amer. Proc. 15: 149-151.

Williams, C.H. and A. Steinbergs. (1959) Soil sulphur fractions as chemical indices of available sulphur in some Australian soils. Aust. J. Agric. Res. 10: 340-352.

4

Determination of Available Phosphorus in Soils

Phosphorus in soils exists in both organic and inorganic forms. Plants absorb phosphorus primarily as HPO_4^{2-} and $H_2PO_4^-$ ions, with the major source of these phosphate ions being inorganic forms. In acidic soils, plants utilize phosphorus from aluminium phosphate, while in calcareous soils, calcium phosphate serves as a source of inorganic phosphorus. All these inorganic forms of phosphorus exist in dynamic equilibrium with water-soluble phosphorus in the soil. Therefore, it is evident that the methods used to determine available phosphorus in soil should focus primarily on these inorganic forms, along with water-soluble phosphorus.

The chemical determination of available phosphorus involves two phases:

1. Extraction of soil

2. Analysis of extract

Numerous attempts have been made to formulate a universal extractant for available phosphorus (P), but so far, none have succeeded. Different extracting solutions have been proposed, and the amount of phosphorus dissolved during extraction varies based on both the quantity present in the soil and the solubility of phosphorus in different solvents. The two methods most commonly used today are as follows.

A. 0.03N NH_4F+0.025N HCI (Bray & Kurtz, 1945)

This method has been most successfully used as an index of available P in acid soils. It has been found to give results that are highly correlated with the crop response to phosphate fertilization. The underlying principle of this method is that the soil is shaken with the extracting solution, which dissolves the fraction of phosphorus available to plant roots. The extract is then analyzed for soluble phosphorus calorimetrically.

Ammonium fluoride in the extracting solution complexes Al^{3+} and Fe^{3+} ions in acid solutions and the phosphorus is rendered free as IS shown by the following equation:

$$3\ NH_4F + 3HF + AlPO_4 \longrightarrow (NH_4)_3AlF_6$$

Hydrochloric acid dissolves more active calcium phosphate and prevents the precipitation of phosphorus (as calcium phosphate) that has been dissolved by ammonium fluoride.

B. 0.05M $NaHCO_3$ pH 8.5 (Olsen et al. 1954)

This extractant is suitable for extracting available P in neutral, alkaline and calcareous soils. This method is commonly used in many soil testing laboratories in India. Its details are given below:

Apparatus

Conical flasks (150 -ml) 2 nos, Balance, Spatula, Pipette (20 ml), Pipette, graduated (10-ml), Funnel, Volumetric flasks (25-ml) 8 nos, Wash bottle, Beakers (100 ml) 2 nos, Electric shaker, Colorimeter.

Reagents

0.5M $NaHCO_3$ solution (pH 8.5): Dissolve 42 g of sodium hydrocarbonate in distilled water and make the volume approximately 950 ml. Adjust the pH of the solution to a pH meter and make the volume to 1 litre with distilled water.

Darco-G 60 (leached with $NaHCO_3$) Solution A: Dissolve 12 g of ammonium molybdate in 250 ml of distilled water and 0.2908 g of antimony potassium tartarate in 100 ml of distilled water. Mix both solutions and add 148 ml of concentrated H_2SO_4 and make the final volume to 2 liters, and store the solution in dark coloured bottle.

Ascorbic acid solution (solution B): Dissolve 1.056 g of ascorbic acid in 200 ml of solution A.

5N H_2SO_4- Dilute 148 ml concentrated H_2SO_4 to 1 litre.

Filter paper (Whatman No.1)

Standard solution of P (50 ppm): Dissolve 0. 2195 g of potassium dihydrogen phosphate (KH_2PO_4) in distilled water and dilute to 1 liter.

Five ppm solution of P: Dilute the 50-ppm solution of P by 10 times.

Principle

Soil is shaken with an extractant at a ratio of 1:20 for half an hour, using Darco-G 60, which absorbs dispersed organic matter in the test sample, resulting in a clearer extract. The extract is then obtained by filtering the suspension through the Whatman No. 1 filter paper. Next, the phosphorus in the extract is treated with ammonium molybdate, a complexing agent, which forms a hetero-poly complex known as phosphomolybdate, characterized by its yellow colour. This phosphomolybdate is then reduced with ascorbic acid, a reducing agent. During this reduction, some molybdenum (Mo) is converted to Mo^{2+} and/or Mo^{4+}, resulting in the complex's characteristic blue colour. The intensity of this blue colour is measured using a colorimeter at a wavelength of 760 nm, utilizing a red filter. The extractant used in this method decreases the activity of calcium by precipitating it as calcium carbonate ($CaCO_3$) in neutral, alkaline, and calcareous soils, thereby increasing the concentration of phosphorus (P) in the solution.

Reactions

Phosphate / Soil Colloid + Bicarbonate → Bicarbonate / Soil colloid + Phosphate

$H_2PO_4^- + 12\ H_2MoO_4 \longrightarrow H_2P\ (MO_3O_{10})_4 + 12\ H_2O$

(Phosphate) Phosphomolybdate (Yellow-coloured)

Phosphomolybdate + Ascorbic acid ⟶ Phosphormolybdate
(Yellow coloured) Reduction (Blue-coloured)

Procedure

Weigh 1 g of soil in a 150 ml conical flask. Add a pinch of Darco—G 60 and 20 ml of 0.5N $NaHCO_3$. Shake the flask for half an hour on an electric shaker, and then filter the suspension through Whatman No.1 filter paper. Similarly, prepare a blank. Pipette 5 ml of the extract in a 25 ml volumetric flask. Add 0.5 ml 5N H_2SO_4 and shake for a while till CO_2 evolution disappears. Add 4 ml of ascorbic acid solution (solution B) to it and make the volume with distilled water and mix the contents of the flask. Measure the intensity of the blue colour developed with a colorimeter at a wavelength of 760 mu, using a red filter. Locate the reading given by the colorimeter in the standard curve and calculate the results.

Preparation of standard curve of P

Pipette into 11 different 25-ml volumetric flasks, 0.5, 0.5, 1.0, 1.5, 2.0, 2.5, 3.0, 3.5, 4.0, 4.5 and 5.0 ml of 5 ppm solution of P respectively. Add 5 ml of 0.5 N $NaHCO_3$ into every flask and 0.5 ml 5N H_2SO_4 and shake for a while till CO_2 evolution disappears then follow the same procedure for colour development as given above. Plot a curve between the concentration of P and colour intensity. If this curve is to be plotted between concentration and absorbance, it can be plotted on any ordinary graph paper but if per cent transmittance is the unit for colour intensity, then the curve can be plotted on a semi-log graph paper only.

Precautions

1. Allow sufficient time for the instrument to warm up before use.
2. Clean the outside of the test tube with filter paper to remove any solution droplets that may be stuck to its sides.

3. When inserting the test tube into the cuvette, ensure that the white line on the test tube aligns with the red line on the cuvette.
4. Each time a new sample is tested, rinse the test tube with the same sample before filling it and inserting it into the cuvette.
5. Always handle the test tube by the top.
6. The filtrate should be colourless.
7. Prepare the ascorbic acid solution fresh for each use.
8. If the colour intensity is too high, use a smaller amount of filtrate to redevelop the colour; do not dilute the coloured solution.

Calculations

Weight of soil taken = 1 g

The volume of extractant added = 20 ml

First dilution = 20 times

Vol. of the extractant taken for colour development = 5 ml

Final volume made = 25 ml

Second dilution = 5 times

Total dilution = 20 x 5 = 100 times

% Transmittance as read from colorimeter = T

ppm of P as read from standard curve against % T = A

ppm of P in the given soil = A x 100 = B

kg/ha of P in the given soil = B x 2.24

Interpretation

Amount of available (kg/ha)	P Rating
12	Low
12-22	Medium
>22	High

References

Bray, Roger, H, and Kurtz, L.T. (1945). Determination of total, organic and available forms of phosphorus in soils. Soil Sci. 59. 39-45.

Black, C.A. (1965). Methods of soil analysis Part 2. Chemical and Microbiological properties. Pp. 866-78, 1040-48.

Jackson, M.L. (1967). Soil Chemical Analysis. pp 134-165,429-451

Olsen, S.R., Cole, C.V. Watanabe,F.S. and Dean, L.A. (1954). Estimation of available phosphorus by extraction with sodium bicarbonate, U.S.D.A. Circular 939-19.

Willard, H.H., Meerit, Jr. L.L. and Dean, J.A. (1965). Instrumental Methods of Analysis. Pp 5480.

5

Determination of Exchangeable Cations in Soils

Introduction

The exchangeable forms of cations are the most important source of immediately available nutrients for plants. This property of soil colloids enables soil to support crop growth by regulating the supply of nutrients to plants. When a plant absorbs nutrients from the soil solution, the nutrient concentration in the solution decreases. Consequently, cations are replenished from exchange sites, ensuring a steady supply of nutrients.

In neutral soils, the major exchangeable cations are calcium (Ca^{2+}), magnesium (Mg^{2+}), potassium (K^+), and sodium (Na^+), which together make up approximately 80% calcium, 15% magnesium, and 5% ($Na^+ + K^+$) of the total cation exchange capacity (CEC) of the soil. Other cations, such as zinc (Zn^{2+}), manganese (Mn^{2+}), copper (Cu^{2+}), and iron (Fe^{2+}), are held in the soil by stronger adsorption forces on the surfaces of mineral grains or form chelates with humic acid. As a result, these cations are usually found in trace amounts on cation exchange sites. The total amount of these basic cations, expressed as a percentage of the CEC, is per cent base saturation (%BS).

$$\% \text{ BS} = \frac{(\text{Ca+Mg+ K+Na})}{\text{CEC}} \times 100$$

Principle

A known-weight soil is repeatedly washed with a neutral normal solution of ammonium acetate (1N NH_4OAc, pH 7). The excess NH_4^+ ions in the solution are adsorbed onto the exchange complex, replacing the cations already bound to it. The cations released into the soil solution, such as calcium (Ca), magnesium (Mg), potassium (K), and sodium (Na), can then be determined using appropriate methods for each cation.

Reaction

$$CH_3COONH_4 \longleftrightarrow CH_3COO^- + NH_4^+$$

$$\text{X-Soil} + NH_4^+ \longleftrightarrow NH_4^+ \text{ - Soil} + X$$

Reagents

1. *Ammonium Acetate Solution (pH 7):* Dissolve 77.09 g of ammonium acetate in water and dilute the solution to a total volume of 1 litre.
2. *Standard Potassium Solution (1000 ppm):* Dissolve 1.907 g of potassium chloride (KCl) in distilled water and dilute it to 1 litre using 1N ammonium acetate solution.
3. *Standard Sodium Solution (1000 ppm):* Dissolve 2.543 g of sodium chloride (NaCl) in distilled water and dilute it to 1 litre using 1N ammonium acetate solution.
4. *0.01N EDTA Solution:* Dissolve 2.00 g of disodium dihydrogen ethylene diamine tetra acetic acid in water, bringing the volume to 900 ml. The normality of this solution should be determined by titrating a 25 ml portion of the standard 0.01N calcium solution. After titration, dilute the solution to achieve an exact normality of 0.01N and perform a standardization check.
5. *0.01N Calcium Solution*: Weigh 0.50 g of calcium carbonate (A.R. Grade) and transfer it to a 100 ml beaker. Add 6N hydrochloric acid (HCl) dropwise until the calcium carbonate is fully dissolved. Dilute the resulting solution to a total volume of 1 litre.
6. *Ammonium Chloride-Ammonium Hydroxide Buffer:* Dissolve 67.5 g of ammonium chloride in 570 ml of concentrated ammonium hydroxide. Dilute the solution to a total volume of 1 litre.
7. *Erichrome Black T Indicator:* Dissolve 0.5 g of Erichrome Black T and 4.5 g of hydroxylamine hydrochloride in 100 ml of 95% ethanol.

Procedure

Weigh out 4 g of medium or fine-textured soil or 6g of coarse-textured soil and transfer into a 50 ml of the centrifuge tube. Add 33 ml of IN CH_3COONH_4 solution having a pH value of 7.0. Stopper the tube and shake for 5 minutes in a reciprocating shaker. Centrifuge the tube for 10 minutes at about 2000 rpm. Decant the clear supernatant liquid as completely as possible into a 100 ml volumetric flask. Extract the sample in this manner three times, using 33 ml portions of IN CH_3COONH_4 solution, collecting the supernatant each time in the same flask. Make to volume, mix and determine the amounts of various cations in the extract using different methods.

Calculations

Weight of soil taken =5g

Volume of NH_4OAc extract prepared =100 ml

Dilution = 20 times

For exchangeable Na

Flame photometer reading =X

Na^+ concentration from standard curve (ppm) =Y

Na^+ concentration in soil (ppm) = Yx20 =Z

Na^+ concentration in soil (me/100 g soil or = Z/ (23 x10) cmol (p)/kg soil

For exchangeable K

Flame photometer reading =A

K^+ concentration from standard curve (ppm)= B

K^+ concentration in soil (ppm) = Y x20=Z

K^+ concentration in soil (me/100 g soil or =Z/ (39x10) cmol (p)/kg soil

For Calcium and Magnesium

The calcium and magnesium in soil saturation extract can be determined either on a flame photometer by using a calcium/magnesium filter for that element or by using versenate method., But versenate method being more accurate is recommended for soil testing laboratories. In this method, versenate (disodium dihydrogen ethylene diamine tetra acetic acid, EDTA) is titrated against Ca or Ca+Mg. Versenate forms stable complexes with calcium, magnesium and several other ions at different PH values. The interfering ions encountered in this system are Cd, Sn, Cu, Zn, Fe and Mn ions. However, the interference of these interfering ions is checked either by using 1% KCN solution or by adding sodium diethyl dithio carbamate crystals.

For CALCIUM

Principle

A known volume of soil saturation extract is titrated against 0.0IN EDTA solution using ammonium purpurate indicator in the presence of 4N NaOH solution. The endpoint is the change of colour from orange-red to purple at a pH value of 11 when the whole calcium forms a complex with EDTA.

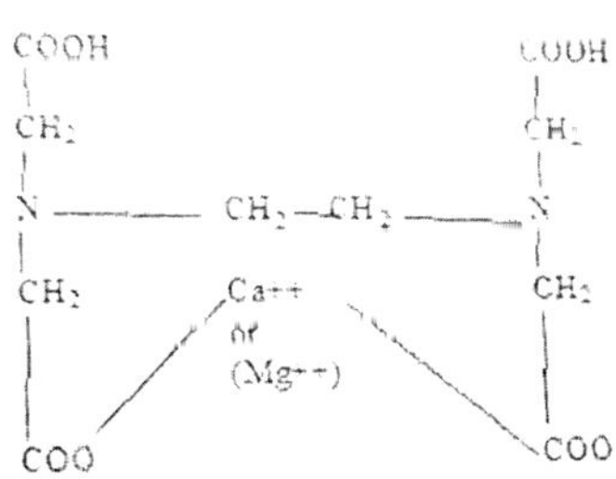

Reagents

1) **0.0IN EDTA solution:** Dissolve 2 g of disodium dihydrogen ethylene diamine tetra acetic acid in water and make volume to 900 ml with water. The normality of this solution is obtained by titration of a 25 ml portion of the standard 0.01N Ca solution, and the solution is then diluted so that the normality is exactly 0.01N. Check standardization is then made.
2) **0.0IN calcium solution**: Weigh 0.50 g of calcium carbonate (A.R. Grade) and transfer it to a 100 ml beaker. Now add 6N HCI dropwise until $CaCO_3$ is dissolved. Dilute the solution to 1 litre.
3) Ammonium purpurate indicator: Thoroughly mix 0.5 g of ammonium purpurate (Muroxide) with 100 g of potassium sulphate (A.R Grade).
4) 4N Sodium hydroxide: Take 160 g of pure sodium hydroxide, dissolve it in water, and make the solution 1 litre.
5) Sodium diethyl dithio carbamate crystals.

Procedure

Pipette out S ml of soil solution in a china dish. Dilute it with 20 ml of distilled water. Add 2-3 crystals of carbamate and 10 drops of 4N NaOH solution. Add approximately 50 mg of purpurate indicator. Stir with the help of a glass rod and titrate with 0.01N versenate solution until the colour gradually changes from orange-red to violet (purple). EDTA solution should be added dropwise, and the endpoint must be compared with a blank titration.

Calculations

Vol. of 0.01 N EDTA solution used =X ml

Milli equivalents (me) of Ca^{++}/ liter = (X × 0.01 x 1000)/ ml of soil saturation extract taken

For CALCIUM

Principle

Magnesium in a given sample of soil saturation extract can be determined by titrating it against a 0.01N versenate solution using Erichrome black T as an indicator and ammonium chloride-ammonium hydroxide buffer. The endpoint will be clear by changing colour from wine red to blue or green at a pH of 10. As calcium is also present in the soil saturation extract, this titration will determine both calcium and magnesium.

Reagents

1) 0.01N EDTA solution
2) Ammonium chloride-ammonium hydroxide buffer: Dissolve 67.5 g of

ammonium chloride in 570 ml of concentrated ammonium hydroxide and dilute to 1 liter.

3) Erichrome black T indicator: Dissolve 0.5 g of Erichrome black T and 4.5 g of hydroxylamine hydrochloride in 100 ml of 95 per cent ethanol.

Procedure

Pipette out 5 ml of soil solution in a china dish. Dilute it with 20 ml of distilled water. Add 2-3 carbamate crystals and two drops of Erichrome black T indicator. Now add 10 drops of NH_4 CI—NH_4 OH buffer and titrate it against 0.01N EDTA solution. The endpoint will be the change of colour from wine red to blue or green. Take at least three concordant readings.

Calculations

Volume of 0.01 N EDTA solution= Y ml

Milliequivalents of (Ca^{++}+Mg^{++})/ liter = (Y x 001 x 10000/ ml of soil saturation extract taken

Milliequivalents of Mg^{++} per liter = (Ca^{++}+Mg^{++}) – Ca^{++} , both in me/L

6

Determination of Micronutrients in Plants

The analysis of plant samples for total micronutrient content involves two main steps. First, the plant tissue is either ashed or digested to eliminate the organic components. Next, the levels of zinc (Zn), iron (Fe), copper (Cu), and manganese (Mn) in the resulting digested extract are measured using an atomic absorption spectrophotometer.

Apparatus/Equipment

Analytical balance, measuring cylinder (25 ml capacity), hot plate with a temperature control system, conical flasks (100/150 ml capacity), polyethylene/glass funnels, polyethylene bottles (60/100 ml capacity), measuring flasks (50/100 ml capacity), pipettes, atomic absorption spectrophotometer, hollow cathode lamps of Zn, Cu, Fe, and Mn, and Whatman No. 1 filter papers.

Reagents

Digestion mixture: A triple acid mixture comprising nitric acid (HNO_3), perchloric acid ($HClO_4$) and sulphuric acid (H_2SO_4) in 9:3:1 or a di-acid mixture of HNO_3 and $HClO_4$ in 4:1 ratio.

Procedure

A. Digestion of plant samples

Plant material can be digested using a triacid or diacid mixture, or through ashing. The wet diacid digestion method, which involves a combination of nitric acid (HNO_3) and perchloric acid ($HClO_4$), is the preferred approach for digesting plant samples. Triacid digestion, which consists of HNO_3, sulphuric acid (H_2SO_4), and perchloric acid in a ratio of 9:4:1, along with digestion using H_2SO_4 and hydrogen peroxide (H_2O_2), is generally avoided. This method is not recommended because H_2SO_4 can introduce micronutrients and heavy metals into the samples. It is also crucial to perform blank digestion, in which all the same steps are followed without including the plant material, to avoid any contamination from the reagents. The following steps outline the procedure for digesting plant samples:

1. Weigh an appropriate amount of the well-processed sample and transfer it into a conical flask.
2. Add 10-15 mL of the diacid mixture and gently mix the contents.
3. Allow the mixture to stay overnight.
4. The following day, heat the flask on a hot plate at a low temperature. This will initiate a vigorous reaction, accompanied by the release of brown fumes of nitrogen oxides (NOx). Once the nitric acid (HNO_3) has boiled off, white fumes of perchloric acid ($HClO_4$) will start to appear, and the reaction will complete within 1-2 minutes.
5. Continue the digestion until the contents of the flask become colourless.
6. Once digestion is complete, allow the flask to cool. Then, dilute the contents with deionized water and adjust the final volume to 50 ml.
7. Filter the solution through Whatman No. 1 filter paper and store the filtrate in a clean polyethylene bottle for further analysis.

B. Determination of Zn, Fe, Cu and Mn in plant digest

The concentration of Zn, Fe, Cu, and Mn in plant digest can be determined using an atomic absorption spectrophotometer, as described in determining DTPA extractable Zn, Mn, Fe and Cu in soil.

Calculation

Weight of plant material taken for digestion = W g

Final volume made after digestion = 50 ml

Dilution = 50/W = A

The concentration of micronutrient as given by AAS = B µg/ml

Content of micronutrient in plant sample = B×A µg/g

Precautions

1. The acids used to prepare the distillation mixture should be of analytical reagent (AR) or guaranteed reagent (GR) grade. Distilling nitric acid (HNO_3) before use is preferable.
2. To ensure complete digestion, it is crucial to maintain a low temperature during the initial stages of digestion and gradually increase it to full heat as the process continues.
3. If the contents of the digestion flask exhibit charring, indicating incomplete digestion, add 2-3 ml of concentrated HNO_3 and continue digestion until the mixture is free of charred material.
4. If the digest contains colloidal material, it should be filtered.
5. A blank digestion should also be performed to account for any impurities present in the reagents used.

6. Deionized water should be utilized for diluting the digests.
7. The instrument must be calibrated to zero using the blank digest, which should be prepared similarly to the plant digest.
8. Lastly, all apparatus used during the analysis must be thoroughly washed and decontaminated.

7

Determination of Total Sulphur in Plant Samples

Sulphur is an essential nutrient for plant growth. While it has traditionally been considered a secondary nutrient, it is increasingly recognized as the "fourth macronutrient," along with nitrogen, phosphorus, and potassium. Oilseeds, legumes, forages, and certain vegetable crops require significant amounts of sulphur. In many crops, the sulphur content is comparable to that of phosphorus. Sulphur plays several crucial roles in plants. It is a vital component in the development of chlorophyll and is necessary for protein synthesis. Additionally, sulphur is required by the rhizobia bacteria found in legumes for nitrogen fixation. It is also a major constituent of one of the enzymes needed for the formation of the chlorophyll molecule. The sulphur content in agriculturally important crops ranges from 0.2% to 3.0%.

Principle

During the digestion of plant samples, plant sulphur is oxidized to sulphate. When barium chloride ($BaCl_2$) is added, it reacts with the sulphate to form a barium sulphate ($BaSO_4$) precipitate. This precipitate creates turbidity in the solution, which is proportional to the amount of sulphate present. Measuring this turbidity allows us to determine the sulphur content of the solution in question. This method is based on turbidimetry (Bardsley and Lancaster, 1960). This turbidity is directly proportional to its concentration in the sample.

$$P - SO_4 + BaCl_2 \longrightarrow BaSO_4 + P - Cl_2$$

Reagents

1. Barium chloride dehydrates ($BaCl_2.2H_2O$): 20-60 mesh crystals, analytical grade. Pass the $BaCl_2.2H_2O$ crystals through a 20-mesh sieve and retain them on a 60 mesh sieve.
2. Gum acacia solution: Dissolve 0.5 g powdered gum acacia in a mixture containing 50 ml glacial acetic acid and 50 ml deionised/distilled water. Store the solution in a refrigerator to avoid the growth of microorganisms.
3. Salt buffer solution: Dissolve 50 g $MgCl_2.6H_2O_3$, 4.1 g KNO_3, and ml ethanol in a 1-litre volumetric flask and increase the volume to the mark.

4. 6 N HCl: It is approximately 1:1 acid: water.
5. Standard S solution: Dissolve 0.5434 g K_2SO_4 (AR grade) in 500 ml distilled water and increase the volume to 1 litre (thus, a 100 ppm SO_4^{-2}-S solution).

Procedure

A. Digestion

The wet diacid digestion method is the most preferred method for digesting plant samples. The procedure is the same as given in the P determination of plant samples chapter.

B. Determination of S in plant digest

1. Take 0, 10, 20, 30, 40, 50 ml of 100 ppm SO_4^{2-}-S solution in 250 ml volumetric flasks.
2. Add 25 ml salt buffer solution to each flask.
3. Fill the volume with deionized/distilled water. The solutions provide 0, 4, 8, 12, 16, and 20 ppm SO_4^{2-}-S , respectively.
4. Transfer 10 ml of the above solution to a 50 ml Erlenmeyer flask.
5. Add 1 ml of 6 N HCl and 1 ml of 0.5% gum acacia solution.
6. Mix the contents by swirling and add 0.5 g $BaCl_2.2H_2O$ crystals.
7. Allow the flask to stand for a minute and then swirl the contents gently until the $BaCl_2.2H_2O$ crystals are dissolved.
8. Read the turbidity on a colorimeter using a blue filter or on a spectrophotometer using 420 nm wavelengths.
9. Plot the absorbance or transmittance against the concentration and draw a standard curve.
10. Repeat steps 4-9 with 10 ml of the sample's di-acid digest and take the reading.

Calculations

S in plant tissue (%) = Reading from the standard curve × 10^{-4}[100/sample weight (g) 50/10(dilution factor)]

References

Bardsley, C.E. and Lancaster, J.D. 1960. Determination of reserve sulphur and soluble sulphates in soils. Soil Science Society of America Proceedings 24: 265-268.

8

Determination of Total Nitrogen in Plant samples

Several factors influence the variation in nitrogen levels, including the top sequence of the land, soil type, soil moisture regimes, and soil properties such as texture, pH, and fertility status. Other important considerations are the nature of the fertilizer used, the type of crop being cultivated, and the nitrogen requirements of that crop. One common determination made during soil fertility assessments is the total nitrogen content in plant samples. The method frequently employed for this determination is known as Kjeldahl's method.

Principle

A known weight of plant material is digested with concentrated sulphuric acid (H_2SO_4) in the presence of a catalyst mixture, which consists of copper sulphate ($CuSO_4$), potassium sulphate (K_2SO_4), selenium (Se), and mercury (II) oxide (HgO). During this digestion, the plant material is oxidized by sulphuric acid, converting nitrogen into ammonium sulphate $(NH_4)_2SO_4$. Copper sulphate serves as a catalyst, facilitating the breakdown of organic matter and the conversion of nitrogen to ammonium sulphate. Potassium sulphate regulates the boiling point of sulphuric acid during digestion, ensuring effective processing. Selenium powder shortens the digestion time needed to achieve maximum nitrogen recovery, while mercury (II) oxide helps prevent nitrogen loss that could occur with selenium alone as a catalyst.

From the resulting ammonium sulphate, nitrogen is determined by distilling the ammonia using a 45% sodium hydroxide (NaOH) solution and a few granulated zinc pieces (or glass beads) to promote smooth boiling. The evolved ammonia is then absorbed in a known excess volume of standard acid (0.02 N H_2SO_4), and the excess acid is quantified through titration with a standard alkali (0.02 N NaOH), using methyl red as an indicator. The endpoint of the titration is indicated by a colour change from pink to yellow.

Apparatus and Reagents

1. Agate pestle and mortar, micro-Kjeldahl distillation assembly, pipettes of variable sizes, 150-ml conical flasks, burette, Kjeldahl digestion flask, and a heater with a regulator.

2. Concentrated sulphuric acid (H_2SO_4).
3. Catalyst mixture: Combine 20 g of $CuSO_2 \cdot 6H_2O$, 1 g of selenium powder, and 3 g of HgO, grinding them together in an agate mortar. Mix this blend thoroughly with 480 g of potassium sulphate.
4. Sodium hydroxide (NaOH) solution (45%).
5. Sulphuric acid (H_2SO_4), 0.02 N.
6. Sodium hydroxide (NaOH), 0.02 N.
7. Methyl red indicator (0.5% in absolute alcohol).

Reactions

i) During digestion:

$2NH_2\text{-}R\text{-}COOH + H_2SO_4 \rightarrow (NH_4)_2SO_4$

ii) During distillation:

$(NH_4)_2SO_4 + 2NaOH \longrightarrow Na_2SO_4 + 2NH_3 + H_2O$

$2NH_2 + H_2SO_4 \longrightarrow (NH_4)_2SO_4$

iii) During titration:

$H_2SO_4 + 2NaOH \longrightarrow Na_2SO_4 + 2H_2O$

Procedure

1. Transfer 1 g of prepared plant material wrapped in filter paper to a 300 ml Kjeldahl digestion flask.
2. Add 10 g of the catalyst mixture and 25-30 ml of concentrated sulphuric acid.
3. Gently swirl the flask to mix its contents, careful not to spill the sample onto the sides.
4. Heat the flask on a low heat setting for 20-30 minutes until frothing ceases, then gradually increase the temperature until the acid reaches its boiling point.
5. Continue heating until the organic matter is completely destroyed and the solution turns a clear light yellow or grey colour, ensuring that the acid condensation reaches approximately one-third of the neck of the digestion flask. Cool the flask and fill it up to 100 ml with distilled water.
6. Pipette 10 ml of 0.02 N H_2SO_4 into a 150 ml conical flask, add 2-3 drops of methyl red indicator, and position the flask under the delivery tube of the condenser, ensuring that the tube dips into the acid.
7. Take 5 ml of the aliquot in the distillation flask and connect it to the mouth of the distillation flask.

Pour 25 ml of 45% NaOH into the distillation flask (containing the aliquot) through the funnel attached to the distillation apparatus. Close the funnel tap and begin distillation, ensuring that the water inlet and outlet of the condenser are open and the steam outlet is closed.

Collect about 30 ml of the distillate. To stop distillation, remove the delivery tube from the conical flask and turn off the heat source.

Titrate the excess 0.02 N H_2SO_4 in the conical flask against 0.02 N NaOH. The endpoint is indicated by a colour change from pink to yellow.

Calculations

Wt. of material digested =1 g

Volume made after digestion =100 ml

The volume of aliquot taken for distillation =5 ml

The Volume of 0.02N H_2SO_4 taken in the receiver=X ml

The volume of 0.02N NaOH used for titration of excess 0.02N H_2SO_4= Y ml

Volume of 0.02N H_2SO_4 used for actual absorption of NH_3 =X- Y ml 1 ml of 0.02N H_2SO_4 −0.00028 g of N

% N in the given plant sample=(X-Y) x 0.00028 x 100/5 x100/1

9

Determination of Total Phosphorus and Potassium in Plant Samples

The amount of nutrients absorbed from the soil varies among different crops and crop species. Not all plants have the same nutrient requirements or compete for nutrients similarly.

A. Digestion of Plant Samples

There are two common methods used to destroy the organic compounds in plant samples:

i. **Dry Ashing:** This method involves igniting the plant samples, and the ashing procedure can vary based on the temperature used. A temperature of 500°C for 2 to 4 hours is generally sufficient to digest the plant samples fully. It is important to note that boron can only be determined using dry ashing, as it volatilizes during wet digestion.

ii. **Wet Digestion:** This process involves the oxidation of organic matter in a liquid medium. The oxidation is performed using a triple acid mixture of HNO_3, $HClO_4$, and H_2SO_4, or a di-acid mixture of HNO_3 and $HClO_4$ in a 4:1 ratio.

Reagents

Triple acid mixture: Mix concentrated HNO_3, $HClO_4$ and H_2SO_4 of analytical grade in the ratio of 9:3:1.

Procedure

1. Weigh 1 g of the plant material and transfer it to a 250 ml Erlenmeyer flask.
2. Add 15-20 ml of the triple acid mixture and mix the contents of the flasks by swirling.
3. Place a funnel at the neck of the flask and keep it overnight.
4. Now heat the flask at a low temperature on a hot plate. A vigorous reaction with the evolution of brown nitrous oxide fumes set in immediately. After HNO_3 is boiled off, white fumes of $HClO_4$ start coming out, and the reaction is over within 1-2 minutes

5. Then raise the temperature to the fullest (around 400) so that the refluxing of H_2SO_4 occurs at the base of the neck of the flask.
6. Continue heating at this temperature until the flask's contents appear yellowish green (1-2 ml of the digest is left behind).
7. Cool the contents and dilute with 25 ml of distilled water.
8. Transfer the solution to a 100 ml volumetric flask and make up the volume by washing the Erlenmeyer flask. Filter it through Whatman No.1 filter paper and the filtrate to determine phosphorus and potassium.

Reference

Piper, C.S. Soil and Plant Analysis.

B. Determination of total phosphorus in the plant extract

Total phosphorus in plant extract is generally determined by vanado molybdo phosphoric yellow colour method in a nitric acid system.

Principle

Vanadate (V^{5+}), molybdate (Mo^{6+}), and orthophosphate (P^{5+}) react to give a yellow complex in acid solutions. The exact nature of the yellow chromogen of the vanadomolybdophosphoric system is not known. Still, the colour is attributed to the substitution of oxovanadium and oxy molybdenum radicals for the oxygen of PO_4^{-3} to give a chromogenic heteropoly compound.

Reagents

i) 4N Na_2CO_3
ii) 6NHCl
iii) 2,4 dinitrophenol indicator
iv) HNO_3 - Vanadate - molybdate reagent
   Solution A is prepared by dissolution of 25 g of ammonium molybdate $(NH_4)_6 Mo_7O_{24}4H_20$ in 400 ml of distilled water. Then, solution B is prepared by dissolving 1.25 g of ammonium metavanadate (NH_4VO_3) in 300 ml of boiling water. Solution B is cooled, and then 250 ml of concentrated nitric acid is added, and the solution is again cooled to room temperature. Finally, solution A is poured into solution B, and the mixture is diluted to 1 litre.
v) Phosphate standard: Dissolve 0.2195 g of KH_2PO_4 dried at 40°C and dilute to 1 litre after adding 25 ml of 7N H_2SO_4. This solution contains 50 ppm P.

Procedure

Take 5 ml of the plant digest in a 25 ml volumetric flask. Add 1-2 drops of 2-4 dinitrophenol indicator and 4N Na_2CO_3 solution dropwise till yellow colour appears. Now add 6 N HCI dropwise till the yellow colour disappears. Add 2.0 ml of 6 N HCI in excess to get the required pH of 4.8. At this stage, add 5 ml vanadate molybdate reagent and make up the volume to 25 ml. The colour develops in several minutes (usually 30 minutes is allowed for full-colour development) and is stable for 2 months at high concentrations but at a P concentration of 5 ppm. It is stable for only 2 weeks. Similarly, prepare a blank. Read the intensity of the yellow colour formed on a spectronic-20 photoelectric colorimeter at a wavelength of 470 nm.

Preparation of standard curve

Pipette out 0, 0.5, 10, 1.5, 2.0, 2.5, 3.0, 3.5, 4.0, 4.5, and 5.0 ml of 50 ppm solution in eleven 25 ml volumetric flasks. Add 5 ml of the extract prepared in the blank to each flask and proceed with the colour development in the same way as in the procedure described above.

Calculations

Weight of plant material taken =1 g

Volume made = 100 ml

First dilution = 100 times

The volume taken for the development of colour=5 ml

Final volume made =25 ml

Second dilution=5 times

Total dilution=100 × 5=500 times

Transmittance as read from colorimeter=T

ppm as read from the standard curve against T value=Y

ppm of P in the given plant sample=Y× 500 =A

(%) P in the given plant sample=$A/10^4$

C. Determination of total potassium in the plant extract

Nutrients in soils exist in various forms, and their availability to crops differs. Some nutrient pools may not be included in standard soil tests, but they can still contribute significant amounts of nutrients during crop growth. For example, sandy soils rich in mica can provide adequate potassium (K) even if soil tests show low available K content. K deficiency is more commonly observed in red and lateritic soils. In contrast, illite-dominant alluvial soils have high potassium reserves and can release sufficient K throughout the crop growth period. However, continuous cropping on these soils without replenishment can lead to the emergence of K deficiency. Total potassium concentrations in plants serve as a reliable indicator of K availability.

Procedure: Transfer 1 ml of the plant extract to a 50 ml volumetric flask. Add distilled water to bring the total volume up to 50 ml. Next, feed this solution into the atomizer assembly of the flame photometer, ensuring that the galvanometer has already been calibrated using standard potassium (K) solutions. Finally, record the reading from the galvanometer.

Preparation of standard curve

To prepare a 50-ppm working solution of potassium (K) from a 100-ppm stock solution, follow these steps:

1. Pipette 0, 2.0, 4.0, 6.0, 8.0, and 10.0 ml of the 50 ppm K solution into separate 50 ml volumetric flasks.
2. Add 1 ml of a blank extract to each flask.
3. Fill each flask to the 50 ml mark with solvent and mix the contents thoroughly.
4. Adjust the galvanometer reading of the flame photometer using a ten ppm K solution.
5. Measure the readings for the other prepared standard solutions.
6. Finally, plot the flame photometer readings against the K concentrations on regular graph paper to create a standard curve.

Calculations

Wt of plant material taken =g

Volume made=100 ml

Ist dilution=100 times

Second dilution=50 times

Total dilution=100 x 50= 5000 times

Reading shown by flame photometer=A

ppm of K against A as read from the standard curve= Y

ppm of K in the plant sample= Y x5000=B

(%) K in the given plant sample=B/10^4

References

Chapman, H.D. and P.F. Pratt. 1961. Methods of analysis for soils, plants and waters, University of California, Division of Agricultural Sciences, U.S.A.

Jackson, M.L. 1987. Soil Chemical Analysis, Prentice Hall of India, New Delhi

10

Determination of Organic Carbon in Soils

Purpose: To assess the fertility level of soil

Carbon is the primary component of soil organic matter, and the estimation of organic matter is based on its organic carbon content. Soil organic carbon values can be converted to total organic matter by multiplying the organic carbon by a factor of 1.724 (Van Bemmelen), which assumes that organic matter, on average, contains 58% organic carbon.

Soil organic matter, which comprises the remains of plants, animals, and microbes in various stages of decomposition, plays a vital role in soil health and functionality. It serves as a storehouse for essential nutrients such as nitrogen, phosphorus, sulphur, and boron and typically accounts for at least half of the soil's cation exchange capacity releasing carbon dioxide. It provides energy and building blocks for microbes, crucial for various biochemical processes within the soil. It also contributes to soil structure stability, thereby influencing moisture retention and aeration. Organic matter, which contains approximately 5% nitrogen on average, is often used as an index of a soil's nitrogen-supplying capacity. Many researchers consider soil's organic matter content to be a crucial indicator of soil fertility.

Reagents

IN Potassium dichromate ($K_2Cr_20_7$) solution: Dissolve 49.04 g. of $K_2Cr_20_7$, in distilled water and dilute to 1 litre.

N/2 Ferrous Ammonium sulphate Fe $(NH_4)_2$, $(SO_4)_2.6H_20$ solution (Mohr's salt): Dissolve 392 g of ferrous ammonium sulphate in distilled water. Add 15 ml of concentrated sulphuric acid (H_2S0_4) of AR grade and make volume to 2 litres with distilled water.

Sodium fluoride powder or orthophospheric acid (95%)

Concentrated sulphuric acid H_2S0_4; 36 N (A.R. grade).

Diphenylamine indicator: Dissolve 0.5 g of diphenylamine in a mixture of 100 ml of concentrated H_2S0_4 and 20 ml of distilled water.

Methods

The soil organic matter is estimated from the organic carbon by the following methods:

1. Dry combustion method

2. Wet combustion method

1. Dry combustion method

The gravimetric method described involves treating soil with sulphurous acid (H_2SO_3) to eliminate calcium carbonate ($CaCO_3$). After treatment, the soil is ignited in silica tubes, which release carbon dioxide (CO_2). This CO_2 is then absorbed by a weighed soda-lime tube, and the amount of CO_2 produced is determined by measuring the difference between the initial and final weights of the soda-lime tube.

From the quantity of soil samples taken and the amount of CO_2 evolved, the content of organic carbon and, consequently, the organic matter can be calculated. Although this method is more accurate, it is also tedious and time-consuming. Additionally, the removal of carbonates can be challenging, and a very finely ground sample is required. Because of these factors, only a limited number of samples can be analyzed in a day.

2. Wet combustion method

This is a rapid and relatively good method. During the day, a large number of samples can be analyzed. Here, the presence of $CaCO_3$ does not affect the determination. The most commonly used wet combustion method is the Walkley and Black (1934) Rapid Titration method. The details of which are given below:

Principle

Oxidizable organic matter in a soil sample is oxidized using potassium dichromate ($K_2Cr_2O_7$). The reaction is facilitated by the heat generated when concentrated sulphuric acid (H_2SO_4) is mixed with a 1 N $K_2Cr_2O_7$ solution. The excess $K_2Cr_2O_7$ is then determined through titration with an N/2 ferrous ammonium sulphate solution, using diphenylamine indicator and sodium fluoride (NaF) to ensure a clear solution due to its flocculating effect. The amount of organic matter oxidized is calculated from the quantity of $K_2Cr_2O_7$ that is reduced or consumed during the oxidation process. The highest temperature achieved from the heat of the dilution reaction upon adding H_2SO_4 is approximately 120 °C, which is sufficient to oxidize the more reactive forms of soil organic carbon but not the more inert forms that may be present.

This method oxidizes a lower percentage of the total organic carbon in the soil. However, it has been found that modifying the method to include external heat immediately after the addition of H_2SO_4 to the sample allows for nearly complete oxidation of organic carbon, resulting in a more consistent degree of oxidation compared to the regular method. In the standard approach, only about 76% of organic carbon is oxidized. Therefore, a correction factor should be used when calculating the percentage of organic carbon in the final results.

Reactions

Oxidation reaction

$K_2Cr_20_7 + H_2SO_4 \longrightarrow K_2SO_4 + Cr_2(SO_4)_3 + 4H_2O + 3O \times 2$

$C + 2O = CO_2$

$2\,K_2Cr_20_7 + 8\,H_2SO_4 + 3C \longrightarrow 2\,K_2SO_4 + 2\,Cr_2(SO_4)_3 + 8H_2O + 3CO_2$

Titration reaction

$FeSO_4(NH_4)2SO_4.6H_2O \longrightarrow FeSO_4 + (NH_4)2SO_4 + 6H_2O \times 2$

$2\,FeSO_4 + H_2SO_4 + O \longrightarrow Fe_2(SO_4)_3 + H_2O$

$2\,FeSO_4(NH_4)_2SO_4.6H_2O + H_2SO_4 + O \longrightarrow 2(NH_4)2SO_4 + Fe_2(SO_4)_3 + 13H_2O$

Behaviour of Diphenylamine Indicator

$$2(C_6H_5.NH.C_6H_5) \xrightarrow[-H_2O]{+O} 2(C_6H_5.NH.C_6H_4) \xrightarrow[-H_2O]{+O} C_6H_5N\text{-}C_6H_4.C_6H_4N\text{-}C_6H_5$$

Diphenylamine	Diphenylamine Benzidine (Colourless)	Diphenylamine Benzidine (Violet)

Procedure

Weigh 2 g of soil into a 250-ml conical flask. Add 10 ml of 1 N potassium chromate ($K_2Cr_2O_7$) solution and mix thoroughly. Next, cautiously add 20 ml of concentrated sulphuric acid (H_2SO_4) while swirling the flask to ensure proper mixing. Leave the flask as such so that the contents can cool and the reaction can complete. Afterwards, add approximately 2 g of sodium fluoride (NaF) powder or 10 ml of orthophosphoric acid (H_3PO_4) as flocculating agents, along with 100 ml of distilled water. Shake the mixture vigorously. Then, add 10 drops of diphenylamine indicator and some ammonium sulphate solution. You will observe a colour change, indicating the end point of the titration—from violet to bright green.

Proceed to titrate the contents of the flask with N/2 ferrous ammonium sulphate solution. Note the volume of ferrous ammonium sulphate solution used for the titration, and calculate the results accordingly. Additionally, perform a blank titration in a similar manner for comparison.

Calculations

Weight of soil taken = Sg

Vol. of N/2 FAS solution used for the blank titration = x ml

Vol. of N/2 FAS solution used for titration of the excess $K_2Cr_2O_7$ = y ml

Vol. of 1 N $K_2Cr_2O_7$ used for the oxidation of carbon =(x=y)/2 ml

Iml of 1 N $K_2Cr_2O_7$ = 0.003 g of organic carbon

% organic carbon in the soil =(x=y)/2 ml x 0.003 x 100/S=A

% organic carbon in the soil = Ax100/76 = A x 100/76 = A x 1.31 =B

% organic matter = B x 1.724

Interpretation

% organic carbon	Rating
Below 0.40	Low
0.40-0.75	Medium
Above 0.75	High

Precautions

1. Add $K_2Cr_2O_7$ very carefully without touching the neck of the flask.
2. Be careful while adding sulphuric acid, as it can spoil your clothes.
3. Read the upper meniscus of the ferrous ammonium sulphate solution.
4. Do not place the flask on a wet surface as it will break.
5. If the contents of the flask turn green with the addition of indicator but before titration, repeat the sample with double the amount of standard $K_2Cr_2O_7$ solution and sulphuric acid.

References

Black, C.A. (1965) Methods of soil analysis. Part 2. Chemical and Microbiological Properties. Pp. 1367-1378.

Jackson, M.L, (1967) Soil Chemical Analysis. Pp. 205-226.

Walkley, A. and Black, C.A. (1934) An examination of the Degtjareff method for determining soil organic matter and a proposed modification of the chromic acid titration method., Soil Sci. 37:27-38

11

Introduction to The Principles and Uses of Analytical Instruments

Measurement is essential in various aspects of life, whether in the kitchen, businesses, research stations, clinical settings, or industries. It plays a crucial role in our daily lives, as many activities would become challenging without measurement. We commonly measure time, length, weight, temperature, air pressure, flow, level, density, and chemical composition, among other parameters, using various measuring devices. Measurement simplifies our lives and directs our daily activities and livelihoods.

Specific devices known as measuring instruments are essential to measuring these parameters. The science of measurement and control is referred to as instrumentation. Analytical instruments precisely measure and control various physical and chemical properties such as pH, conductivity, density, viscosity, octane number, and chemical composition. Instruments used in laboratories are often referred to as Bench Top Laboratory Analyzers. These analytical instruments can be complex and require intensive maintenance.

Analytical instruments allow us to determine the chemical compositions of substances by answering questions like: "What elements are present? What compounds exist? In what quantities are these present?" This process aids in gathering both qualitative and quantitative information about the matter being tested.

Qualitative analysis often employs physical properties such as boiling points, melting points, solubility, refractive indices, and colour reactions, while quantitative analysis utilizes methods like gravimetry, volumetry, and titrimetry. This distinction is crucial because even minor impurities can significantly alter a substance's characteristics.

The instrumental methods used for the measurement of various physical and chemical properties are given below:

1. Optical Principle Analytical Techniques
    - Absorption of Radiation: Photometry, Spectrophotometry, Atomic Absorption

- Emission of Radiation: Flame Photometry, Emission Spectroscopy, Fluorescence
- Scattering of Radiation: Turbidimetry, Nephelometry
- Diffraction of Radiation: X-ray diffraction, Electron Diffraction

2. Electrochemical Methods
   - Measurement of Electrode Potential: Potentiometry
   - Measurement of Mass Deposited on Electrode: Electrogravimetry
   - Change in Conductivity: Conductimetry
   - Change in Diffusion Current concerning Potential: Voltammetry/ Polarography
3. Mass Change Determination:
   - Mass Spectrometry
4. Radioactivity Measurements:
   - Radioanalytical Methods

We will discuss a few analytical instruments, including the spectrophotometer, atomic absorption spectroscopy, and flame photometer.

An introduction to these analytical instruments and their principles follows below.

Colorimetry

Colorimetry is an analytical method used to determine the concentration of a coloured solution by comparing it with a known standard. This process involves evaluating solutions of coloured substances, typically dissolved in colourless solvents. The comparison is based on the intensity of the colour in the unknown solution versus that of a solution with a known concentration (the standard) in the same solvent. The principles of colorimetric analysis are based on the Lambert-Beer Law, which can be expressed as follows:

$A= \log_{10} I_o/I= kcb$

Where A is absorption,

I_0 is the intensity of the incidental radiation,

I is the intensity of the transmitted radiation, k is a proportionally constant known as "Molar Extinction Coefficient' c is the concentration of absorbing species, b is the thickness of the absorbing medium.

The ratio $1/I_0$ is called the transmittance, which is related to absorption as shown by the following equation:

$A= \log 100/T=2- \log T$

The measurement of absorption or transmittance is performed using an instrument called a colorimeter, which operates based on the above-given principle. Many colorimeters are available today, utilizing photocells to measure colour intensity. All colorimeters consist of the following main components:

1. Light Source: Emits radiation at different wavelengths.
2. Filter: Selects the radiation of the desired wavelength.
3. Sample Holder: Contains a test tube with the tested solution.
4. Photocell: Converts light energy into electrical energy.
5. Galvanometer: Measures the light intensity (converted into electrical energy) in terms of optical density (absorbance) or transmittance.

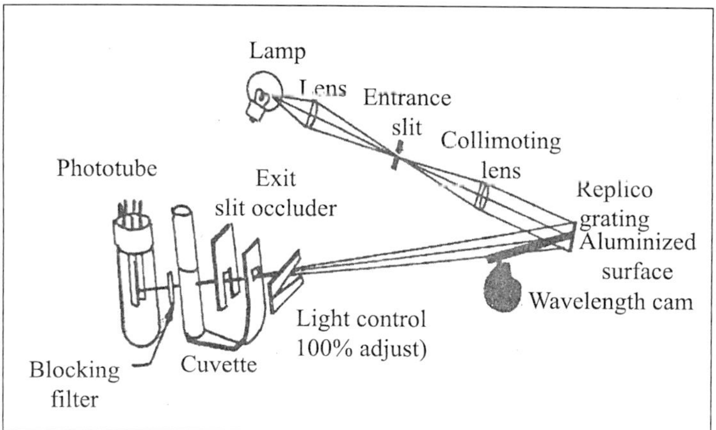

Figure: Schematic optical diagram of Bausch & Lomb Spectronic 20 calorimeter

Calibration of the Colorimeter

1. Plug the instrument into a 230 V AC power source at 50 Hz (single phase).
2. Insert the filter with the required wavelength into the designated slot.
3. Turn on the instrument and allow it to warm up for 15 minutes.
4. Fill a test tube with the blank solution and insert it into the sample holder in the test position (i.e., cuvette). Adjust the meter needle to zero optical density (100% transmittance) using the "Increase Light" knob. After setting it to zero, place this test tube in the additional space in the sample holder for occasional rechecking.
5. Rinse the previous test tube with the test solution, then fill another test tube with the test solution. Note the absorbance or per cent transmittance as indicated by the galvanometer needle.

Atomic Absorption Spectroscopy

The technique in which the absorption of radiation of a characteristic wavelength by the atomic vapours is related to the concentration of an element is called atomic absorption spectroscopy. The instrument using these techniques is known as the Atomic Absorption Spectrophotometer (AAS) and is used to estimate micronutrients in the test solutions. It was first used as an analytical technique, and the underlying principles were established by Robert Wilhelm Bunsen and Gustav Robert Kirchhoff, both professors at the University of Heidelberg, Germany, in the second half of the 19th century on their study on sodium spectrum.

Principle of Atomic Absorption Spectrophotometer

Atomic absorption spectrophotometer is based on the principle that when atomic vapours of an element are irradiated by the radiation from a light source (i.e. hollow cathode lamp filled with neon or argon at low pressure) whose emission lines are those of the element in question, they absorb it in direct proportion to the concentration of the element being determined in the test sample.

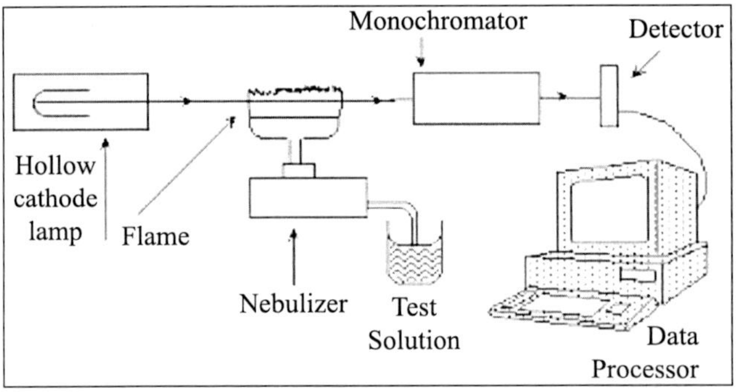

Figure: Schematic diagram of Atomic Absorption Spectrophotometer (AAS)

Instrument features

Though a wide range of atomic absorption spectrophotometers (AAS) are available these days, yet all of them have the following common basic components:

i) Light source: It emits the spectrum characteristic of the element to be determined. The most widely used light source is a hollow cathode lamp, which is designed and operated so that the lines to be measured are sharp and have a stable intensity.

ii) Atomizer-burner assembly: It helps convert the test solution into atomic vapours. The solution to be analyzed is drawn up by a capillary and converted by means of a stream of compressed air to a fine spray. After condensation of larger droplets, the fine spray is mixed with the fuel gas acetylene (C_2H_2) and burnt in a long flame (at 2100-2400°C) in a stainless-steel burner.

Index

iii) Monochromator: It isolates the absorbing resonance line from other nonabsorbing lines. It performs its function when the light coming from a hollow cathode lamp, after traversing the flame, enters the monochromator, which is already set at the wavelength of the resonance lines of the desired element.

iv) Detector: It measures the magnitude of absorption of the characteristic radiation.

v) Photomultiplier tube (PMT): It amplifies the absorption signal and converts the light radiation into electrical energy measured in optical density or absorbance.

vi) Readout System: It measures the absorbance in volts, usually by using a strip chart recorder, a digital display, a meter, or a printer.

Calibration

1. Prepare the stock standard solutions containing 1000 mg of the given micronutrient per litre by dissolving the required quantity of their respective salt (for Zn-4.398g of $ZnSO_4.7H_2O$; Cu-3.929g of $CuSO_4.5H_2O$; Fe-4.977g of $FeSO_4.7H_2O$ and for Mn-3.598 g of $MnSO_4.H_2O$) and making the final volume to one litre with deionized water.
2. From 1000 mg l^{-1} stock solution of Zn, Cu, Fe and Mn, prepare 50 mg l^{-1} working solution by pipetting out 5 ml of 1000 mg l^{-1} solution in a 100-ml volumetric flask and makeup the volume with the deionized water. From this solution. Prepare the required standard solutions of each element in 50-ml volumetric flasks by pipetting out the required volume of the working standard and making the final volume with 0.005 M DTPA. The concentration ranges usually required are 0, 0.2, 0.4, 0.8 and 1.2 mg l^{-1} for Zn; 0 0.2, 0.4,0.8, 1.0 mg l^{-1} for Cu; 0, 2, 4,6,8, and 10 mg l^{-1} for Fe and 0, 1,2,3, and 4 mg I^{-1} for Mn.
3. Now feed blank to set the instrument's zero. Then, the standards of the desired element are fed one by one to standardize the instrument to read the absorbance A and/or concentration (C) of the unknown samples within a particular range of concentration. Prepare the standard curve by taking the concentration of the element on the X-axis and absorbance readings on the Y-axis.

Flame Photometery

It is a fact that the atoms of an element when excited by some heat or electrical energy, emit radiations of a characteristic wavelength. The technique by which the intensity of such characteristic radiations of an element is correlated with the concentration of that element is known as flame photometry or flame emission spectrometry. When the liquid-gas aerosol enters the flame, the solvent evaporates, leaving behind the solid-gas aerosol. The solid particles of the aerosol are further vaporized in the flame to gaseous molecules, which, on further heating, dissociate into neutral atoms. As the energy of atoms is subjected to changes in level by absorption of light energy, therefore, when the neutral atoms are further excited, the electrons shift to positions more distant from the atomic nucleus. When these electrons regain their stable state (i.e. when the excited atoms return to lower energy levels), radiation enters the dispersing device in the instrument to isolate the desired spectrum region. A photocell and some meter measure the intensity of the isolated radiation, directly proportional to the amount of the element emitting the radiation.

Flame photometry is particularly valuable in determining the alkali and alkaline earth elements. These emission spectrophotometric methods are very useful when a large number of soil and plant analyses have to be made routinely. The instrument used for emission spectrophotometric analysis is known as a Flame Photometer, which consists essentially of the following components:

1. Pressure regulators and flow meters for the fuel gas: Generally, a 10-lb gauge for the fuel gas and a 25-lb gauge for the oxygen or air supply are employed.
2. An atomizer device associated with the flame source: This component consists of the following parts:
   i) Suction pipe: through which the sample is sucked up.
   ii) Drain pipe: through which the heavy particles of the sample are drained out.
   iii) Air pipe: which is connected to the air compressor and sprays the sample into the flame.
   iv) Fuel gas pipe: It is connected to the gas (a mixture of acetylene and hydrogen) cylinder.
   v) Burner: where the fuel gas burns to give a flame into which the sample is sprayed.
3. An optical system proceeding appropriate photosensitive detectors: It consists of
   i) Filter: It selects the radiation emitted by the element to be determined.
   ii) Photocell: It converts the light energy (i.e. emitted radiation) into electrical energy.

An electrical circuit for measuring the intensity of the radiation: It consists of:

A galvanometer measures the intensity of radiation (which has been converted into electrical energy).

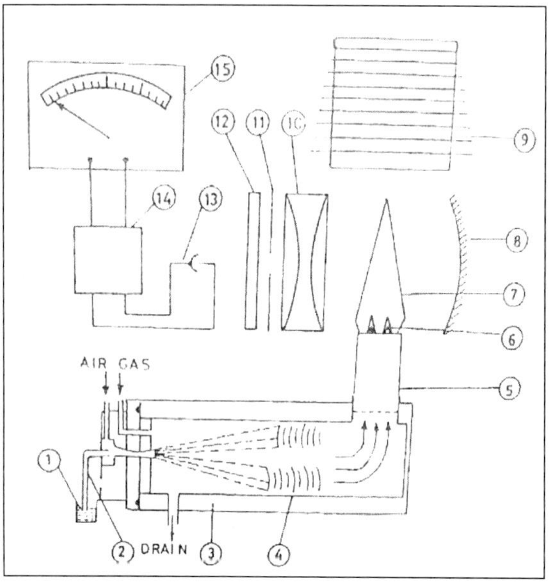

1. Sample Solution, 2. Atomiser Intake Tube, 3. Mixing Chamber, 4. Mixign Vanes, 5. Burner, 6. Inner Cones, 7. Non-Luminous Flame, 8. Reflector, 9. Chimney Top Detachable, 10. Condenser, 11. Aperture, 12. Filter, 13. Photo tube, 14. Amplifer and 15. Meter

Figure: Schematic diagram of Flame photometer

Working and calibration of the Flame Photometer

i) Plug in the instrument to correct voltage.

ii) Insert the filter of the element to be determined in the holder.

iii) Set the air compressor in operation and regulate the air pressure between 0.4 to 0.75 kg/ cm^2.

iv) Turn on the fuel pressure and immediately light the burner with a match stick carefully through the inspection window.

v) Adjust the gas pressure to obtain a soot-free-blue flame (about 4 inch in height). Close the inspection window and let the flame burn for 10 minutes before feeding the standards and sample.

vi) Set the galvanometer needle at zero by feeding distilled water to the atomizer.

vii) Feed the standard solution to the atomizer and set 100 of the galvanometer.

viii) Feed the test sample and note down the reading indicated by the meter needle.

ix) When the job ends, turn off the fuel and air and switch off the galvanometer.